for the House of Lords Library
from the authors.
27.1.82

MONOGRAPHS ON SCIENCE, TECHNOLOGY, AND SOCIETY

THE POLITICS OF CLEAN AIR

Eric Ashby
and
Mary Anderson

CLARENDON PRESS OXFORD
1981

Oxford University Press, Walton Street, Oxford OX2 6DP

OXFORD LONDON GLASGOW
NEW YORK TORONTO MELBOURNE WELLINGTON
KUALA LUMPUR SINGAPORE JAKARTA HONG KONG TOKYO
DELHI BOMBAY CALCUTTA MADRAS KARACHI
NAIROBI DAR ES SALAAM CAPE TOWN

Published in the United States by
Oxford University Press, New York

British Library Cataloguing in Publication Data

Ashby, Eric
The politics of clean air. - (Monographs on science, technology, and society)
1. Air-Pollution - Great Britain
I. Title II. Anderson, Mary III. Series
363.7′392′0941 TD883.7.G7

ISBN 0-19-858330-3

Typeset by DMB (Typesetting), Oxford
Printed in Great Britain by Eric Buckley
Printer to the University
Oxford University Press
Oxford

… our successors will wonder at the ludicrous ingenuity with which we have so long managed to diffuse our caloric and waste our coal in the directions where they least conduce to the purposes of comfort and utility.

2nd Annual Report to the Hon. the Commissioners of Sewers of the City of London, 26 November 1850.
John Simon, Officer of Health to the City

PREFACE

This book is written for general readers interested in history and public affairs. We have, however, drawn upon published and unpublished primary material; so for scholars we have provided an ample list of references to the sources we have used for our story. We advise the general reader to disregard the superior figures in small print scattered through the text, for these refer only to the references to sources.

Messrs Heyden and Son, publishers of the journal *Interdisciplinary Science Reviews*, and Dr Anthony Michaelis, editor of the journal, kindly permitted us to include in Chapters 1, 5, and 7 some material published in another form in that journal. We have pleasure in recording our thanks to them, and especially to Mr P.M. Williams, who encouraged us to expand into a book the essay we wrote for the journal.

We are indebted to several people who have helped us. Mr M.F. Tunnicliffe, a deputy chief inspector of the Alkali Inspectorate, generously lent us material essential for a history of the Inspectorate. The Department of the Environment gave us access to reference 14 in Chapter 10. Mr R.J. Ogden, assistant director in the Meteorological Office, gave us data on the occurrence of fogs from 1841 to 1976. Mr W.P. Quinlan, principal statistician at the British Gas Corporation, gave us data on the sales of gas. Rear Admiral P.G. Sharp, secretary-general of the National Society for Clean Air, Miss E.D. Mercer, head archivist of the Greater London Record Office, and Miss Betty Masters, deputy keeper of the records, Corporation of London, gave us useful help over some of our inquiries. We are grateful also to Mrs Bridget Boehm, who typed the book with great accuracy and astonishing speed.

We wish also to thank the Master and Fellows of Clare College, Cambridge, for a grant toward the cost of typing and copying the manuscript.

EA

February 1981 MA

CONTENTS

1

'TO CONSUME THEIR OWN SMOKE'

Prelude, 1819

'The air we receive at our birth and resign only when we die is the first necessity of our existence.' This resonant plea for clean air appeared in *The Times* on 17 February 1881.[1] It is a fitting introduction to our theme. Unlike other primordial elements of antiquity, earth and water, air cannot be owned even by the State. How, then, can air be protected from pollution and made safe to breathe?

For centuries common law protected individuals from specific nuisances due to polluted air. Thus in 1691, when Thomas Legg of Coleman Street in London complained of the smoke from his neighbour's bakehouse, the baker was ordered to put up a chimney 'soe high as to convey the smoake clear of the topps of the houses'. But common law is made to protect people, not air. If air is to be protected, whether or not it is causing a nuisance, the State has to intervene. This book tells the story of State intervention: the tortuous evolution of a national policy in Britain to control smoke and what the Victorians called 'noxious vapours'.

In the sixteenth century there was a minor energy crisis in Britain. Wood and charcoal for burning were becoming scarce and were being replaced by soft coal. By the middle of the seventeenth century fires and furnaces were producing the 'Hellish and dismall Cloud' over London, vividly described by John Evelyn. Then followed the development of steam power. Coal was used to drive mills, to smelt ores, to brew beer, to make bricks, pottery, glass, and soap. Polluted air, for centuries a private nuisance, became a public menace. It was an affront not just to eyes and nostrils but to health. In London, weekly Bills of Mortality, compiled to record visitations of the plague, were used as long ago as 1662 by John Graunt to establish a correlation between death rates and the burning of coal.

Concern was sufficiently aroused to stimulate invention. Appliances to enable engines to consume their own smoke were devised and patented; James Watt registered one of the earliest patents in 1785. In some districts the police were given powers to compel owners of steam engines to adopt these appliances. But these efforts got nowhere. Industries multiplied and flourished and the air over industrial cities became more sulphurous and foul. At long last, in 1819, the issue was brought to parliament. It is at this point that our story begins, appropriately in Whitehall Gardens.

In Whitehall Gardens lived the MP for Durham City, Michael Angelo

Taylor. It was evidently a less salubrious area than Whitehall is today. This is the sort of thing Taylor had to put up with: 'the volumes of smoke which issue from the furnaces on every side of the river Thames opposite my own house actually blacken every flower I have in my own garden in Whitehall'. Like many small men (the caricaturists stressed his diminutive height) Taylor was energetic and articulate. He had already been responsible for improvements in paving and lighting in the metropolis, and on 8 June 1819, he launched a campaign against smoke. He moved for a select committee

to inquire how far it may be practicable to compel persons using Steam Engines to erect them in a manner less prejudicial to public health and public comfort.[2]

The committee was appointed forthwith and produced its first report in July, before the parliamentary recess.[3] It was revived in the following session (May 1820) and a further report was presented two months later.[4] Unlike many later initiatives in parliament to abate pollution, this initiative had a modest success: the Bill which followed the select committee's reports, though diluted with exemptions, did pass into law on 28 May 1821.[5]

Let us follow the deliberations of the select committee and the debates in the House, for they bring to life the state of public opinion about the polluted air of cities 160 years ago. First, witnesses confirmed what scarcely needed confirmation, that smoke was bedevilling life in many industrial cities: not only London but Manchester, Liverpool, Birmingham, and elsewhere: 'even the other day', said one witness, 'in passing through Great Russell-street, Bloomsbury, I observed the street stifled with smoke issuing from a neighbouring brewery, and extending from Oxford-street, where the brewery is situated, far beyond the British Museum'. Residents in Liverpool, said another witness, used to be able to see right across the town to the harbour twenty years ago. Now, twenty years later, 'the mouth of the harbour can scarcely be seen during the week'.

Already this offensive by-product of the industrial revolution had set the seeds of decay in towns and cities. One of the social consequences of smoke was that people who could afford it no longer lived near their place of work. Taylor made this point vividly, in the face of incredulity on the part of his fellow MPs, in a debate in the House when his committee was reconstituted in May 1820. He was speaking of industrial smoke:

From the manner in which these furnaces were erected, and increased as they were in number, London, Manchester, Liverpool, Birmingham and other extensive cities and towns were almost become uninhabitable. [At this point there was laughter] Gentlemen might laugh at the term, but he contended it was strictly true, as the persons engaged in business in these towns were actually compelled to guard the health of their families from the deleterious effects of the ebullitions of smoke issuing from those furnaces, by removing to hired houses at a distance from the place of their avocation.[6]

One hundred years later, in 1920, the fight for clean air was still going on, and another committee of inquiry was considering how to wage it. Taylor's assertion, which brought laughter in parliament in 1820, was still 'strictly true'. Two members of this later committee went to the industrial area of Rhine-Westphalia to see how Germans abated smoke. They found the smoke-abatement laws so effective that great towns like Dusseldorf were 'pleasant and agreeable places of residence... even the richest citizens continue to live within the city boundaries, a practice which has long since been abandoned in British manufacturing towns'.[7]

The evidence assembled by Taylor's committee in 1820 was enough to persuade any reasonable person that cleaning the air, like cleaning the streets, was a public responsibility and parliament could no longer be indifferent to it. But acts of parliament are useless if they cannot be enforced. Was it possible to abate smoke without halting the progress of industry?

On this question, too, Taylor's committee heard persuasive evidence. The committee's remit covered only steam engines and furnaces (furnaces were added to the terms of reference when the select committee was appointed). This restriction was deliberate. Taylor had consulted all sorts of experts before he raised the matter in parliament and he had learnt that it was practicable to abate smoke from furnaces used to generate steam for these could be stoked at a steady rate; but it was not practicable to abate smoke from furnaces which had to be run on and off in bursts, or which were used for smelting ores or vitrifying glass. So his objective was limited; to cut down smoke from steam engines—though this objective was abundantly worth aiming at, for in the 1820s the main source of air pollution was from the generation of steam.

Taylor's faith, that steam engines could be designed to consume their own smoke, rested on the invention of a worsted manufacturer from Warwickshire, Josiah Parkes. Parkes, anxious to cut down smoke from his family business, constructed furnaces on the plan of James Watt, but he did not find them to be successful. He was inspired to try a different design through his acquaintance with Humphry Davy, from whom he learnt 'what, till then, was not so well known; viz. the true theory of combustion, as then published to the world... in his [Davy's] researches on flame and on the safety lamp'.[8]

The 'true theory of combustion' was simple, though its successful application to the abatement of smoke has taken the better part of a century to work out. In complete combustion, carbon in coal is transformed into colourless, odourless carbon dioxide, by combination with oxygen from the air. The main constituent of smoke is unburnt carbon. Therefore, to persuade a furnace to consume its own smoke all that is necessary is to introduce enough air at a high enough temperature. James Watt was the first of a long line of inventors to grasp the need for an auxiliary supply of air, but—

according to Parkes—he admitted the air at the wrong point.[9] Parkes's improvement on Watt's design was to admit air through an adjustable valve at the back of the furnace. Parkes claimed that 'an hour after lighting the fires we have heat enough to consume the whole of the smoke arising from them'. And, of course, there was a promise of a further benefit: the less smoke, the more coal is burnt, and therefore the less fuel is needed.[10]

Taylor seized on this critical evidence in favour of his case. He promptly travelled to Warwick, in company with another MP, Thomas Denman—a brilliant lawyer who later on helped him to draft the pioneer Bill to control smoke—to see Parkes's furnaces for himself. After rigorous tests and careful observation, Taylor was so impressed that he persuaded Parkes to take a missionary trip to London to prevail on other industrialists to adopt the design in order to demonstrate its efficacy to the public there. Parkes duly waited on Messrs Barclay and Perkins and in May 1820 his appliance was fitted to two steam boilers at their brewery, and also to a steaming copper. This was at about the time that Taylor's select committee was revived. Before it began to take evidence again Taylor arranged a visit to the newly adapted furnaces at Barclays for members of the committee together with other influential persons including members of the Commons and the Lords. His faith was vindicated. 'Every trial', he told the committee, 'was made in my presence, and every trial succeeded; and I have no doubt that if there was occasion, several members of this House, who were present, would attend and give evidence to the same fact.'[11]

Several members of the committee did in fact give such evidence, but the key witnesses were Parkes himself and others who had adapted their furnaces to his plan. Parkes was able to report further success with other boilers, notably two serving a 90 horse-power engine in Birmingham. But he did not overestimate his success. He warned the committee about the limitations of his device: it would not work on furnaces which were constantly having to be started up and damped down.[12] Nevertheless Taylor, armed with a gratifying quantity of evidence, proceeded to frame a Bill to be introduced in the 1820-1821 parliamentary session. His original intention was 'to propose a declaratory law making the present construction [of furnaces] a nuisance'.[13] Fortunately he was deterred from presenting anything so drastic and the Bill he moved in April 1821 was more modest and ingenious—reflecting, no doubt, the legal astuteness of its eminent sponsors (it was supported not only by Denman, who had seen for himself the smoke-free chimneys in Warwick, but also by the redoubtable Henry Brougham, who, with Denman, had defended Queen Caroline during her trial in 1820).[14] In their more expert hands the Bill was framed, not to supersede common law prosecutions for nuisance but to make them marginally more effective. In the event of a conviction for nuisance due to polluted air, the courts might award costs to the prosecutors (not hitherto possible in indictment proceedings); and to prevent such a nuisance recurring, the courts might order any alterations in the furnaces they judged expedient, before passing final sentence. Thanks to Parkes's cautious reservations about the value of his device, the Bill referred solely to furnaces 'employed in the working of engines by steam'. It was prefaced by a bold and confident assertion of the technical grounds on which it was based:

> Whereas great inconvenience has arisen, and a great deal of injury has been and is now sustained by His Majesty's subjects...from the improper construction as well as from the negligent use of furnaces employed in the working of engines by steam; *And whereas it has been ascertained, that the greater proportion if not the whole of the smoke issuing from such Furnaces, may, by a proper construction and due attention, be consumed, so as to prevent the Nuisance thereby arising...* [italics ours]

And then followed the innocuous clauses which were aimed at no more than the encouragement of prosecutions under common law. Innocuous though they were, they were enough to provoke resistance from persons likely to be inconvenienced by them. Taylor found, when the Bill came for debate in the Commons, that some of his brother MPs did not share his confidence and were unconvinced by his evidence. Members from mining and industrial districts questioned whether Parkes's device (or any other) would abate smoke; they doubted whether more efficient combustion would economize fuel; they said (they were right about this) that a lot depended upon the skill with which furnaces were stoked, and skilful stokers were hard to come by; they resented the notion that courts of law might interfere in the design of furnaces; they murmured about the need—if the Bill went through at all—to exempt some counties from its application and to exclude some processes from the threat of prosecution. A member from Staffordshire urged that 'Parliament ought to hesitate before they imposed a compulsory expense and inconvenience on so many persons...Were the measure to be confined to the metropolis, he should not object to it; but there were many parts of the country in which it would be extremely injurious. [He meant, of course, injurious to the factory owner, not to his employees living under a pall of smoke.] In the south of Staffordshire there were above 2,000 steam engines and in the neighbouring counties at least 5,000 more.'[15]

Despite apprehensions, doubts, and pleas for postponement, the Bill went through. Its wings were clipped—it exempted even steam furnaces in mines, for instance—and its brave overture was muted by the excision (in the House of Lords) of the passage printed in italics above. But it was on the statute book.[16] The State had taken a first timid hesitant step toward a policy for clean air: and those whose interests were threatened by a clean air policy had put up their first defence.

There is no way to tell whether Taylor's Bill made any difference to the pall of smoke over England. Parkes, speaking twenty years later, said that it had 'frightened the manufacturers; and for a while it frightened them into the adoption of my plan; the pressure from without (if I may say so) produced some good to me, and to them too...'[17] So locally—and perhaps particularly in London—the new law may have had some visible effect but only as a deterrent; it was not put to the test in the courts, and this meant that the massive problems blocking any State policy for air pollution lay undisclosed and unsuspected.

Parkes's device failed to secure general adoption. It needed only lax control and incompetent stoking to convince manufacturers that the device was no use. Humphry Davy, himself disappointed at the way miners had neglected his safety lamp, summed up the situation in a conversation Parkes recollected:

'You have done right to take out a patent for it...but I doubt very much your introducing it into any general practice.' Parkes asked him why. 'Because it is so simple. You will find the masters too careless, and you will find the men too stupid; and like my lamp, they will not use it though it were to save their lives.'

Meanwhile steam engines multiplied; they were hailed as miraculous agents of civilization: they augmented wealth; wealth would lift the labouring class into the middle class; a new middle class would elevate the level of public standards and morality, which in turn would create enlightened political institutions.[18] The reality, grim and ugly, was a denial of this euphoria. Textile workers from the small dispersed water mills in northern valleys were drawn away from their cottages into crude boom towns, packed into shoddy back-to-back houses, often without any water supply. These hives of workers became breeding grounds for disease. And the new pattern of society was hidden from the sun by a blanket of smoke from innumerable chimneys.

In the absence of any lead from central government some towns introduced a 'smoke clause' into their own Improvement Acts. 'And be it enacted', ran the clause in the Derby Improvement Act 1825, 'that from and after the 1st day of January next, all furnaces employed or to be employed in the working of engines by steam, and all furnaces employed or to be employed in any mill, factory, brewery, bakehouse, gas-works, or other buildings used for the purposes of trade or manufacture within the said borough (although a steam engine be not used or employed therein) shall be constructed in the best manner known or practised so as to consume their own smoke' and there followed the threat that a breach of this law or negligence in the use of any furnace would entail the offender if convicted in a fine of 42 shillings.[19] Other towns passed similar laws, including Leeds, in 1842, where the punishment was also a fine of 42 shillings for offenders who had not used the best practicable means (a formula which runs through the whole of our story) to prevent or abate the nuisance.[20] Quite apart from the trivial penalty, the owner of the furnace was well protected against anyone optimistic enough to bring suit against him. How do you define 'in the best manner known or practised'? How, indeed, do you define smoke? And what are the 'best practicable means'?

Mackinnon's campaign, 1843–1850

Michael Angelo Taylor vowed, in 1820, that he would never cease his efforts until he had compelled 'every town in England' to adopt Parkes's device to cut down smoke,[21] but in fact he made no further move for the thirteen years more that he remained in parliament. The issue for a time ran into an impasse. It was in Manchester that the blockage was removed. Credit for this goes to a controversial and combative Anglican clergyman, the Revd J.E.N. Molesworth, vicar of Rochdale. He led what must be one of the earliest pressure groups against air pollution: The Manchester Association for the Prevention of Smoke. On 13 June 1843 Molesworth submitted a petition to parliament, asking for an inquiry into the prevention of smoke in manufacturing towns, with a view to legislation.[22] It was surely more than a coincidence that Molesworth's brother-in-law, a member of parliament named W.A. Mackinnon, asked parliament a fortnight later to set up just such an inquiry.[23] Parliament agreed to appoint a select committee and Mackinnon was made its chairman.

Taylor had good reason, back in 1819, to appeal to parliament to pass laws to abate smoke. He could not walk in his garden (he complained) because of the smoke pouring across the river from the Lambeth waterworks. Mackinnon's motive was much more altruistic. None of the constituencies that returned him to parliament over a span of thirty years was in an industrial area. He himself was born in 1789—the year of the French Revolution—the eldest son of the chief of the clan Mackinnon in the Western Isles of Scotland; brought up as a member of the landed gentry. But he had a lively and genuine concern for the welfare of folk less fortunate than himself. The sorrows and sufferings of the poor, he wrote, were 'indeed sacred things'.[24] For eight years, from 1843 to 1850, he kept up a campaign for laws to abate smoke.

Mackinnon made it his first task to dispel indifference about smoke. Considering how limited the influence of public opinion was in those days, it is remarkable how seriously he took it. 'In a civilised community', he wrote, 'the form of government and its liberal tendency depend on the state of society, not the state of society on the form of government.'[25] So he set to work vigorously on his select committee to collect evidence. In a couple of months the committee met sixteen times and heard opinions from nine scientists (including Faraday and three other Fellows of the Royal Society), thirteen engineers or designers of furnaces, and seven manufacturers and others. The Committee reported to the House of Commons on 17 August.[26] Some recommendations in the report were challenged when it was debated in parliament; but there was sufficient interest to set up a further select committee, still under Mackinnon's chairmanship, in 1845. The committee took more evidence and issued two more reports.[27] The assembled evidence in these reports gives a clear impression of the state-of-the-art of smoke prevention at that time and began to disclose the difficulties that lay in the way of any effective legislation.

Of prime importance, because it was one of the causes of persistent delays in government action, was the fact that no witness could produce hard evidence that smoke was injurious to health. John Graunt, in the seventeenth century, had shown a correlation between smoke and mortality; but a correlation is no proof of causation. Other factors beside smoke—malnutrition, poverty, long hours of work—obviously exacerbated the symptoms of tuberculosis, bronchitis, and asthma. So although there was much talk about the unhealthiness of smoke, the committee, when it came to making recommendations, had to rely on nuisance value more than on hazard to health. Even the nuisance value was subjective, for although there were hints about the social costs of smoke (e.g. Meux's brewery in Tottenham Court Road testified that they were obliged to use a smokeless coal (anthracite) because 'the gentlefolks in the squares compelled us to do it...they said it [ordinary coal] made so much smoke in the drawing-rooms and injured the furniture')[28] nevertheless no one in the 1840s had thought of putting a price tag on the damage done to buildings and other property.

The case against smoke on grounds that it was a health hazard was not strong enough to underpin the select committee's recommendations, but there was plenty of evidence about the causes of smoke and the cures for it. One expert witness after another explained that smoke was simply coal dust which escaped combustion for one or other, or both, of two reasons: not enough oxygen for combustion or too low a temperature for the carbon to be oxidized into carbon dioxide. If combustion is complete, there is no smoke. If smoke appears, as Faraday said in evidence to the committee, it must 'depend upon the convenience or the ignorance of the user, the manufacturer. In large fires, like those of steam-engines...it depends more, I think, upon his ignorance than his convenience'.[29]

Unburnt smoke was wasted fuel. Better combustion would save fuel. In theory, therefore, it was in the manufacturer's own interests to prevent smoke. But—and this is what became apparent as the committee listened to the witnesses—in practice the technology of efficient combustion was beset with problems. First among these was a human problem: the stoker. Many witnesses agreed that if a boiler was stoked properly, much less smoke came out of the chimney. But stokers were drawn from 'the lowest dregs of society almost', underpaid, unco-operative toward their masters, and difficult to supervise. Even Josiah Parkes, whose device for admitting air from the back of the furnace had been the mainstay for Michael Angelo Taylor's success in the 1820s, admitted to Mackinnon's committee that the stokers were more important than the engineer who looks after the engine. Their carelessness could burn out a boiler; Parkes quoted an example of this where, as he put it, a half crown piece (presumably a bonus on the wage) would have determined the fate of the boiler.[30] Despite the autocracy of the masters and the weakness of trade unions in those days, a suggestion that the stoker himself should be fined if the chimney smoked did not commend itself to the committee. Parliament was already uneasy about strikes, for in August 1842 boilermen in Lancashire had drawn the plugs from boilers in the factories as a protest against a reduction in wages.

The committee was repeatedly assured, however, that a cure was available in the design of boilers and furnaces, and one design by C.W. Williams, was recommended as a proven way to abate smoke. It was an adaptation of Parkes's invention; air was admitted to the furnace in thin jets through numerous holes so that it was evenly distributed over the burning coal, on the principle of the well-known argand lamp. Parkes himself told the committee, with admirable detachment:

> I have no hesitation in saying that I consider Mr. Williams to have made a step beyond me, and an important one, as his mode of admitting air simplifies the management of the furnaces and accommodates it better to the intellect of the firemen.[31]

Williams's design had been tried in practice. It was, said Andrew Ure, professor of chemistry in Anderson's College, Glasgow, 'a "perfect plan" for preventing smoke: so applicable to different furnaces as to be used in 20 great steamers with most perfect results.'[32]

Most of the encouraging evidence which the committee heard was about abatement of smoke from steam boilers. It was acknowledged that even the best designed and managed furnaces smoked when they were starting up—and subsequent legislation has always allowed for this—but the committee was convinced that once a steam boiler was running steadily there was no need for it to emit much smoke. There was a strong case for compelling furnaces used to generate a continuous pressure of steam to consume their own smoke.

There were other sources of smoke and noxious vapours for which there was no practicable means of abatement, either because the furnaces were used intermittently, or because greater power was required, or because the noxious vapours did not come from coal at all. Examples were iron foundries, brick kilns (still a cause of serious air pollution), and factories making pottery or glass. It was not until the select committee's report had been debated in parliament and had recruited considerable sympathetic support that manufacturers began to realize that their interests were at risk and that they had better organize themselves into a lobby to raise objections and difficulties. This they did so successfully that, as we shall see, all attempts to get Bills through parliament were thwarted for years.

There was one other source of smoke which no politician dared to challenge for another fifty years or more. That was the smoke from domestic hearths, the kitchen range and the drawing room fire. By using smokeless fuel (anthracite or coke) in closed stoves, or by adopting the traditional domestic stove to be found in Russia, Scandinavia, and Germany, it would have been possible, technologically, to eliminate much of the smoke from domestic houses. Technologically, perhaps, but certainly not psychologically. The cheery open fire was the focus of family life. Quite apart from the enormous cost and trouble involved, the conversion of the open English fire would have precipitated a social revolution of the kind no politician would contemplate. Besides, there was a common belief that the continental stove was unhealthy and even dangerous, as it well might be if the gaseous products of combustion leaked into the room. One of the distinguished scientific witnesses, Professor Ure, had decided views about this—a nice illustration of the part which prejudice plays in the setting of social norms and practices. Commenting on the advocacy to the select committee, by Dr Reid, of the continental stove, Ure condemned them

> as the greatest of evils. The sallow and withered complexions of the people most subjected to the influence of these stoves, their headaches and dyspeptic ailments, are well known to observant English travellers, who perpetually contrast the foul stagnant air respired in these apartments, with the fresh invigorating atmosphere of an English parlour, as heated by the open cheerful grate.[33]

Mackinnon had made a good start in his task to dispel ignorance and indifference, but he had not yet raised public opinion to a level which persuaded the government to take the initiative to promote a Bill. So in May 1844 he presented a Bill himself 'to prohibit the nuisance of smoke from furnaces or manufacturies'.[34] His Bill referred only to 'such Furnaces as are employed for the heating of steam boilers' and declared it unlawful 'for the occupiers of any Furnace or Chimney to permit opaque Smoke to issue from such Chimney for any longer period of time than is necessary for the kindling of the fire'—and the burden of proof of innocence was to rest with the defendant.

It was a modest enough start, but it soon ran into trouble. All sorts of weakening amendments were introduced; collieries, mines, and mineral undertakings were exempted; and after further hostile speeches the Bill was postponed. But at least the issues had been ventilated. Some manufacturers began to pay more attention to smoke abatement. The public began to awaken to the possibility that a perpetual pall of smoke over industrial towns might not always be the inescapable price to pay for prosperity.

Mackinnon was not the sort of man to be disheartened by this initial defeat. For another six years, from 1845 to 1850, he patiently worked to dispel the apathy of parliament. Altogether six Bills were presented, debated, and abandoned. To get a clean air Act on to the statute book must have seemed to be a Quixotic ambition, doomed to disappointment. But, looked at from the distance of history, every one of these abandoned Bills fulfilled a purpose: they edged parliament closer to a point of decision and they created expectations among the people which in the end had to be satisfied. Let us look briefly at this glacial change of mood.

In 1845, a year after Mackinnon's first Bill had failed, the Health of Towns Commission issued its second report.[35] Its paragraph on 'nuisances' set smoke second only to the 'evils arising from defective drainage and cleansing'. This prompted Mackinnon to bring in a second Bill.[36] It provoked immediate opposition, this time from the redoubtable John Bright, himself a cotton spinner from Rochdale, an owner of steam furnaces. 'The House', Bright said, 'might employ itself much better than in this peddling legislation, which never could be attended with useful and permanent results.'[37] But even John Bright could not strangle the Bill at birth: there was enough interest to move parliament to set up the second select committee referred to on p. 8. In addition to the technical evidence (already summarized) heard by that committee, there was evidence of growing resentment at the very idea that smoke emissions from an ironmaster's chimneys should be subjected to some sort of surveillance from a government official; it would be 'worse', said one ironmaster from Staffordshire in evidence to the committee, 'than one of the plagues of Egypt'.[38]

The de la Beche–Playfair Report

Under the camouflage of defeat, Mackinnon's campaign had in fact scored two minor victories. First, *The Times*—the analogue in those days to modern mass media—took up the cause in an outspoken editorial: 'the evil of smoke has reached a most intolerable height'.[39] Second, the government, though still unwilling to sponsor any Bill for cleaner air, did decide to take expert advice on whether laws to control smoke would be practicable; a noteworthy decision, for it prompted them to appoint two scientific advisers, pioneers in the service of science to politics. They were Henry de la Beche, director of the Geological Survey and Lyon Playfair, chemist to the Geological Survey and professor in the new School of Mines (a forerunner to the Imperial College). Their business was to make two inquiries. The first was to discover whether the anti-smoke clauses which some towns had included in their Improvement Acts (e.g. Leeds, Manchester, Bradford, and Derby) were effective and whether they might be models for national legislation. The second was to examine the claims for exemption from these anti-smoke clauses which were being made on behalf of certain industries.

The de la Beche–Playfair report was published in 1846.[40] The authors dismissed the scientific and technical issues in a few paragraphs. They reasserted what had emerged clearly from the evidence given to Mackinnon's select committees: that if combustion of coal is complete there is virtually no smoke; that to abate smoke is to save fuel; and that failure to abate smoke may be due to bad design of furnace or to careless stoking. They then went on to reveal the real cause of the failure of anti-smoke clauses to control smoke. It was not primarily technological failure: it was legal failure. Under the Derby Improvement Act, for example, a case in court could not be won unless it could be proved that smoke had continued without interruption for a whole week. There were always periods, however brief, when a chimney did not smoke; in any case, smoke could not be observed at night. So, the report concludes, 'no conviction can be obtained under this Act'. Even if this gap in the law could have been sealed, there were other gaps. The penalty for emitting smoke, generally 40 shillings, could not be levied until after 'one month's notice in writing' and a month's notice had to be given also before proceeding to prosecute. So a chronic polluter could not be brought to court more than six times a year and his total fine would not amount to more than £12: a cheap fee for indiscriminate and persistent pollution. It is no wonder that authorities did not bother to take polluters to court.

So the message of the report was that laxity in the law rather than ignorance of technology was the chief impediment to the abatement of smoke in industrial towns. But de la Beche and Playfair were not sanguine about the prospects of tightening the law. Their whole report was hedged around with qualifications and reservations. 'As an abstract statement', they said, smoke could be abated, but they saw 'grave difficulties' in the way of 'a general law to the effect that it shall be unlawful for chimnies, after a certain date, to emit smoke'. It would be feasible to control smoke from furnaces generating steam, but for a great many industrial processes (distilleries, ironworks, glass works, potteries, and the like) it would be either very difficult or even impossible to apply a general law to abate smoke. With studied circumspection—so exasperating to politicians who want scientists to give yes-or-no answers—they asserted on the one hand that the continued emission of smoke was an unnecessary consequence of combustion; and on the other hand concluded that it was a subject of very great difficulty, 'a difficulty...which we have found [to] increase and not diminish during our inquiries'.

No government, presented with a report in that tone of indecision, would be prepared to back a clean air Bill. Questioned about the report at the beginning of May 1846, the Home Secretary, Graham, replied that the report, 'so far from removing his doubts, had confirmed them'.[41] Undismayed, Mackinnon brought in a third Bill, much on the lines of his first one.[42] In August 1846, when the Bill came up for debate, there was at any rate a fresh government to consider it, for the Corn Law crisis had swept away Peel's administration; it was now Lord Morpeth, not Sir James Graham, who acted as spokesman for the matter. He appealed to Mackinnon to withdraw his Bill in response to a pledge that he—Lord Morpeth—would 'see whether it would be practicable to carry out the suggestions of scientific and practicable men, with a view to legislation'.[43] Mackinnon acquiesced; again he had, under the guise of retreat, made a slight advance. 'The best way, perhaps', he said, 'of carrying the subject, was by mooting it continually, so that the pressure might come from without upon any Government that might be in office.' This was just what was happening. Mackinnon was by now getting letters from all over the country urging him to promote his Bill and warning him of the censure he would suffer from the press if he failed to do so.

In February 1848 the government fulfilled the pledge Lord Morpeth had given, though not in the way Mackinnon would have wished. A comprehensive Public Health Bill was introduced which included a clause for the abatement of smoke.[44] But it was to apply only in districts where the Act was adopted, and for existing furnaces its requirements were permissive, not mandatory. All the same, Mackinnon had scored another point; for the first time the government had included smoke abatement in its national programme of legislation. At the same time—it is still 1848—Lord Redesdale introduced a Bill in the House of Lords to control smoke.[45] It was virtually a facsimile of Mackinnon's last Bill. And concurrently with these two Bills, the City of London promoted a private sanitary improvement Bill which contained an anti-smoke clause.

This really was a reward for the six years of campaigning by Mackinnon, to have three Bills in parliament at the same time with provisions for abating smoke. Parliament was now evidently awakened to public opinion. But the immediate effect was, not surprisingly, to rally opposition. The opponents mobilized themselves and went into battle to defend their right to pollute. They scored some significant amendments in the government's Bill, e.g. the addition of the debilitating words 'if practicable' together with a list of processes to be exempted from any prohibition. This was not enough for the militants; they reported that people in the north of England were alarmed by the Bill: it would cause unemployment; it would be meddling with manufactures; it was premature in the light of the present state of knowledge. The government surrendered: the Bill was sent to the House of Lords bereft of its smoke clause.

Their lordships, perhaps because they were less concerned with commerce, took a more sympathetic view. They restored the smoke clause, though they did add some exemptions to the list: bricks, tiles, and lime works. When the Bill came back to the House of Commons its enemies were prepared to renew their attack. Bright spoke first and delivered a caustic criticism of the smoke clause. He had opposed every smoke Bill introduced into the House and this clause, he said, 'contained all the absurdities of all the former measures put together'. The Bill proscribed *opaque* smoke; 'did not everybody see that the opacity of smoke...would...depend very much upon whether there was a black or a white cloud behind it?' And who was to decide what constituted a 'well approved plan' (one of the phrases in the Bill) for the consumption of smoke? In Lancashire, said Bright, 'no three men were ever found to agree upon any effectual plan for preventing smoke'.[46] Mackinnon did his best to reaffirm the convictions of his select committee, but he was too urbane and polite a man to scotch Bright's sarcasm. The smoke clause was dropped, and with it dropped also the measures proposed by Lord Redesdale and by the City of London.

Notwithstanding the defeat of the fourth Bill presented since 1844, two more were introduced, one in 1849 and another in 1850; both in the House of Lords.[47] By now the industrial lobby in parliament was well organized and the crusaders for clean air were tired. Mackinnon resigned sponsorship of the last Bill to another member of parliament. The defenders of *laissez faire* were so sure of themselves that Bright, though he was in the House for the debate, did not bother to speak. Both Bills were dropped.

Six Bills to bring in a modest measure of smoke abatement for the whole country had failed. Only one feeble encouragement reached the statute book. In 1847 a Towns Improvement Clauses Bill was passed; its intention was simply to harmonize local legislation for such towns as cared to promote improvements. It included a model smoke clause,[48] requiring every furnace in any manufactory to be so constructed as to consume the fuel it used. The clause did not even include the escape phrase 'if practical' but it slipped through parliament unchallenged. It was, after all, only a permissive clause, not a mandatory one; time enough to oppose it if any town had the courage to try to use it.

The evil of smoke had reached what *The Times* called 'an intolerable height'. The Health of Towns Commission had indicted smoke as a nuisance almost as bad as defective sewage. Majorities in parliament had voted in favour of some of the Bills at all but their decisive last stages. What were the lessons to be learnt from the experience of environmental pioneers in the 1840s?

First, that the time was not ripe (a valuable cliché in this context) for state intervention. No one had *proved* that smoke was a hazard to health. No one had invented the ideal device for abating smoke; proof of this was the trickle of fresh inventions—three or four a year—purporting to improve upon the devices of Parkes and Williams. No one had defined opaqueness in smoke or the best practicable means of abating smoke in ways which would ensure successful prosecutions in court. A wise legislature does not pass laws that cannot be enforced, and there was not much prospect of enforcing laws about smoke when the defendants would almost inevitably be industrial Goliaths (possibly themselves sitting on the magistrates' bench) and the plaintiffs would be private Davids, unable to brief clever lawyers. Mackinnon's campaign had aroused the resentment of the 'chimney aristocrats'. They were incensed that a country gentleman should advocate that they, creators of material prosperity for the nation, should submit their industries to state intervention. Mr Foster, MP for Berwick on Tweed, put it tersely in one of the debates: 'There was no end to such kind of legislation, and, if persevered in, there might, in time, be a Bill to prevent expectoration in the streets.'[49]

Another lesson to be learnt from this episode of the 1840s was that smoke was still too low in the priorities for public concern. The 1840s were not a good time to promote what would have been regarded as inessential, though desirable, reforms. Britain was recovering from a trade depression. It was no time to fiddle about with new designs of machinery 'in the present state of trade, when every shilling is an object to the manufacturers, competing as they are with the Continent', as a witness to Mackinnon's committee put it.[50] Between 1845 and 1848 governments had more pressing problems on their plate than the abatement of smoke. There were all the other, and worse, effects of uncontrolled urbanization; there was tragedy in Ireland; and in 1848 there was a frightening outbreak of cholera.

But Mackinnon's strategy was on the right lines: government does depend upon the state of society, not the state of society upon government. His years of patient pressure for the control of smoke had raised a groundswell of public opinion which carried the campaign into the next decade. In the 1840s, for the first time, the problem of smoke had been put under the lens of public scrutiny. In the 1850s parliament began to respond.

Palmerston acts

In history, as in science, it is common to attribute particular achievements to the efforts of particular men or women but it is rash to assume that if the particular men or women had not been on the spot at the time, the achievements would not have been made. Yet it does seem reasonable to conclude that without Mackinnon's sustained efforts parliament would not have been sufficiently softened up to agree to legislation which followed in the 1850s. And without the efforts of two other men the pressure for legislation might have been delayed even beyond the 1850s. These two men were John Simon, medical officer of health to the City of London, and Lord Palmerston who for a brief period (1852-53) became Home Secretary.

John Simon's major preoccupation was with sanitary reform but he interpreted this to cover much more than sewage and drains. His annual report to the Commissioners of Sewers in 1850 carried a powerful and eloquent plea for prompt action to abate London's smoke. Smoke, wrote Simon, indicated 'mismanagement and waste...all the smoke that hangs over us is wasted fuel'.[51] This was the right line to take with Victorian business men. Simon's advice was adopted. A smoke clause was written into the new City of London Sewers Bill, which received the royal assent in July 1851. In other cities smoke clauses had become a dead letter—if indeed the letter had ever been living at all—partly because authorities were unwilling to prosecute and partly because there was no practicable means of policing the law. Simon saw to it that this would not happen in the City of London. Over 115 compulsory notices were served on offenders in the first year and defaulters were strictly prosecuted.[52] This was possible because the City of London is a small area and Simon's authority in it was at that time unchallenged.

But the City is little more than a large parish embedded in a great conurbation. Smoke was no respecter of the City's boundaries and Simon's next step was to urge that similar smoke clauses should cover the whole of the metropolis. In June 1852 the Sheriff from the City of London attended at the bar of the Commons to present a petition: 'praying that similar Provisions to those contained in the City of London Sewers Act, 1851, for abolishing the nuisance of Smoke from Steam Engines and Furnaces may be extended to the whole of the Metropolis'.[53]

There the petition might have rested had not Palmerston, by a happy coincidence, come to preside at the Home Office six months later. Barred from the Foreign Office, owing to his recent high-handed conduct of foreign affairs, he was appointed in December 1852 to the Home Office, and for the next couple of years brought his galvanic powers to bear on domestic issues before being swept into the office of Prime Minister. Palmerston had already demonstrated his sympathy over social problems; he had voted in favour of the Ten Hour Bill and he had worked with determination to suppress the slave trade; moreover he had a family link with Lord Shaftesbury who, as Lord Ashley, had been a potent political force for social reform in the early part of the nineteenth century. The portfolio of the Home Office was a temporary censure for Palmerston, but he brought to the office his impulsive energy and his impatience at any kind of inaction. Fortunately for the cause of smoke abatement, this is what soon engaged his attention. In July 1853 he issued this terse minute:

Prepare Bill enacting that from and after the 30 Novr next Every Furnace within the Metropolitan District, and Every Steam Boat on Thames between London Bridge & Richmond Bridge shall be fitted with an apparatus for consuming Smoke or shall burn Coke instead of Coal. Penalty twenty Pounds for first Conviction and Ten Pounds for Every Day offence is continued after first Conviction.
Query what Court should adjudge & are these Penalties suitable?

P 18/7-53 [54]

Ten days later such a Bill was presented for first reading in the House of Commons.[55] Perhaps because it was confined to the metropolis of London it did not bring down the wrath of the northern industrialists, but it had no easy ride either. It was watered down by the addition of two qualifications which promised relief to polluters. There was a clause saying that 'consume or burn the Smoke' should not in all cases be held to mean 'consume or burn all the Smoke'; also that Justices might remit penalties if they were of the opinion that the defendant had 'so constructed or altered his Furnace as to consume or burn as far as possible all the Smoke arising from such Furnace and has carefully attended to the same, and consumed and burned as far as possible the Smoke arising from such Furnace'. Despite these dilutions the Bill was severely criticized as a measure 'impossible to carry out', one which would cause 'great expense to the manufacturers of London' and produce, besides, 'a total disarrangement of the northern coal trade'. Mackinnon himself spoke, scoring a point against his adversaries by saying that the 'whole objection arose from a disinclination to incur expense in altering furnaces, and from the coal owners of the north, who were apprehensive that anthracite coal would supersede the consumption of theirs on the London market'.[56]

Palmerston's anger against the industrialists who had ganged up against his Bill erupted in a splended denunciation:

> If ever there was a case in which, he would not say the interests, but the prejudices, of the few were opposed to the interests of the many, this was such a case. Here were a few, perhaps a 100 gentlemen, connected with these different furnaces in London, who wished to make 2,000,000 of their fellow inhabitants swallow the smoke which they could not themselves consume, and who thereby helped to deface all our architectural monuments, and to impose the greatest inconvenience and injury upon the lower class. Here were the prejudices and ignorance, the affected ignorance, of a small combination of men, set up against the material interest, the physical enjoyment, the health and the comfort of upwards of 2,000,000 of their fellow men. He would not believe that Parliament would back these smoke producing monopolists, and he was ready therefore with great confidence to go to a division.[57]

His confidence was justified by 66 votes to 12, notwithstanding that the debate was in the early hours of the morning. Thereafter the Bill had an uneventful passage and on 20 August became the Smoke Nuisance Abatement (Metropolis) Act, 1853.[58]

It had been a long haul since Michael Angelo Taylor carried his Bill into law thirty-two years earlier, and at first sight it seems that not much distance had been covered. But Palmerston's Act, unlike Taylor's, was explicitly drawn in order to rid the air of a nuisance, not just to make prosecutions for nuisance easier under common law. Nevertheless, the new Act might have drifted into the backwaters of unfulfilled legislation if it had not been followed up with Palmerston's Napoleonic passion for detail. Two weeks after the Act was passed Palmerston was peppering his civil servants with minutes. 'Has Mr. Minton answered the Inquiry about his Smoke consuming Furnace arrangts.'[59] And later: 'Let me have a List of all the Persons who have *made Inventions* for consuming or preventing smoke in the last Twelve months.'[60] By December 1853 he was obviously contemplating some extension of the law, for he asked the municipal officers of all towns with smoke-abatement clauses to report on how effective the clauses were and 'if those Provisions have not been carried into effect I should wish to know what have been the obstacles which have stood in the way'.[61]

Shortly after the day the Act came into effect Palmerston discovered that although the offences and penalties were clearly set out, no arrangements had been made to administer it—to police the law and to bring culprits to court. He was alerted to this oversight, it seems, by a leader in *The Times*, complaining that nothing was being done to enforce the law.[62] Two days after this rebuke Palmerston minuted: 'It is Time that some organised arrangt should be made for carrying into Effect the Smoke act of last year and I should wish the Commissr of Police to suggest some arrangt for the Purpose.'[63] Hard upon this came volleys of minutes from him urging prosecutions: 'what we want is an Eye Survey and a Report of some one great offender...One or Two Convictions would soon set the Rest on the alert...The way to succeed is to begin...'[64] A couple of weeks later he was asking: 'have any of the River Steamers been found smoking against Law and have any Steps been taken as to any other offenders'.[65] He did not get a clear reply, and promptly stiffened his request: 'Let me have a Report as to whether the River Steamers do or do not smoke'.[66]

This spurred the police into action. By 11 September 1854, he was told, there were some 6500 cases under investigation.[67] Palmerston's next problem was the laxity of the magistrates. Only four cases had been proceeded with; three of these were adjourned and the fourth offender fined £5 with 12 shillings costs. 'These', minuted Palmerston, 'are very unsatisfactory Results. When a Case is proved the Fine ought to be levied & no adjournment...Say this to Magistrate & that I expect him in future to do his Duty. better [*sic*]'. Palmerston's under-secretary felt obliged to remind his master that it would not be proper for him to give orders to a magistrate; so it was agreed that the Crown counsel should press for the full fine in every case.[68]

It soon got abroad that the Home Secretary was being tough in the administration of his Smoke Act. In the eight months from August 1854 to March 1855 (shortly after Palmerston was called from the Home Office to become Prime Minister, to tackle the crisis in the Crimea) there were 147 prosecutions, 124 convictions, only 10 cases dismissed or withdrawn because the nuisance was immediately abated, and only 13 cases adjourned or awaiting hearing. The total of fines exceeded £500.[69] This harvest must have had some effect on the atmosphere of the metropolis.

The public conscience was stirring over the need to combat pollution of the air. It is significant that even during Palmerston's brief tenure at the Home Office criticisms of the Act were directed to stiffening it rather than relaxing it. The focus of criticism was the criterion for exemption from its penalties, namely that the polluter could be excused if he could demonstrate that he had used the best practicable means to abate his smoke. The real issue (wrote the Board of Health, in a minute sent to the Home Office in November 1853) was whether smoke was causing injury or annoyance; the danger now was that arguments in court would neglect this issue and concentrate upon whether or not the defendant had used the best practicable means to avoid causing injury or annoyance. Experts could always be found, by plaintiff or defendant, who would differ in defining what were best practicable means; magistrates would in these circumstances be inclined to give the defendant the benefit of the doubt.[70] This formula, so common that it came to be abbreviated to 'b.p.m.' and persists to this day, plays an important part in our story; it began as a great obstacle to the enforcement of clean air laws, it evolved into an indispensable prescription for their effective enforcement.

2

'THE MONSTER NUISANCE OF ALL'

Lord Derby and noxious vapours

Opaque smoke, despite John Bright's gibe about the difficulty of defining it, was the visible—even tangible—evidence of the unacceptable face of industry. But anyone who lived near potteries, brickworks, or chemical works was aware of invisible noxious vapours, often much more of a nuisance than smoke. No laws were made specifically to control these because little was known about them; virtually the only recourse open to people plagued by such vapours as fluorides from brickworks or acids from soda works was through the common law.[1] Pollution from noxious vapours was more local than pollution from smoke but it was often much more virulent; in particular there was appalling devastation of the countryside from the emissions of hydrochloric acid (then called muriatic acid) from alkali works in parts of Lancashire, Northumberland, and Lanarkshire.

Since the 1830s the alkali industry had grown enormously. Its product was sodium carbonate, used in the manufacture of soap, glass, and textiles. In the process, hundreds of tons of hydrochloric acid were thrown into the atmosphere every day. An article in *The Times* in 1862 describes the effect of the emissions: 'Whole tracts of country, once as fertile as the fields of Devonshire, have been swept by deadly blights till they are as barren as the shores of the Dead Sea.'[2] Polluters could not plead that they had used the best practicable means to cure the nuisance, for as early as 1836 William Gossage had devised a simple and inexpensive treatment. All one had to do was to wash the ascending gases with a descending stream of water, for water absorbs 48 times its own volume of hydrochloric acid. Alkali manufacturers knew this well enough. Some of them had even installed the necessary equipment, but without the restraint of law they had no incentive to see that it was properly operated, and they were unwilling to take the trouble unless all their competitors did the same. There were endless complaints from farmers and landowners, but prosecutions were rarely successful because the victim could not prove from which of the many chimneys the acid had come that had spoilt his crops or woodlands. Providentially it happened that Lord Derby's own estate was in one of the blighted areas, five miles or so east of the alkali works at St Helens; and the estate of Sir Robert Gerard—a staunch supporter of Lord Derby's party—was also in a blighted area. For motives not altogether disinterested, and encouraged by a deputation which waited on him in 1862, Lord Derby took up the matter in the House of Lords. On 9 May 1862 he proposed the appointment of a select committee 'to inquire into the Injury resulting from noxious Vapours evolved in certain manufacturing Processes, and into the State of the Law relating thereto'.[3]

It was a delicate matter to raise and a delicate moment to raise it. Derby's party was not in power—he was leader of the Opposition in the House of Lords—and England's textile trade was in a grave crisis, for the American civil war threatened to cut off four-fifths of the supply of raw cotton for the mills of Lancashire. But Derby was an accomplished speaker ('the Rupert of debate', he was called) and he represented himself as a loyal son of Lancashire, his home county, unwilling to place 'the slightest check' on industrial development:

> If at any time it was important that no undue interference should take place, it would be so at a moment when a great portion of our manufacturers in the north were suffering under the deepest depression, and when the diminution of manufacturing employment would be a more than ordinary calamity. He was the last person who would desire to interfere with the increase of our manufacturing prosperity, if only for the reason that half of his property was situated in the manufacturing districts...

To this Derby added a disclaimer that he was starting 'any course so extravagant as a crusade against the smoke nuisance' such as Palmerston had imposed upon the metropolis in 1853. He wanted the select committee to concentrate its attention on the alkali industry and the damage it was causing to wealthy landowners and poor tenantry alike. He emphasized that for this one noxious vapour there was already 'a most perfect and complete remedy' and that the only obstacle to the use of the remedy was the lack of any law to impose its use upon manufacturers. Local boards could, under the Public Health Acts, prevent the erection of new polluting works, but not all districts had local boards and where they did exist they were often dominated by local industrialists. Derby was careful not to make this a landowner *versus* factory-owner issue: he wanted the select committee to ascertain 'whether legislative measures could be introduced on this subject not only without injury, but with profit to our manufacturers'.

The committee was appointed three days later under Derby's chairmanship, and four days after its appointment the committee began to hear witnesses. By 18 July 1862 the report was completed and laid before both Houses.[4] It was the first time parliament had received advice on noxious vapours, as it had earlier received advice on smoke. The chief scientific witness was again Lyon Playfair. His opinion was less reserved and hesitant than it had been over smoke in 1846.[5] He thought that noxious vapours from the manufacture of soda, sulphuric acid, and ammonia salts, were all 'within what has been ascertained practically to be removable' and thus within the compass of legislation. But 'the monster nuisance of all' was hydrochloric acid from alkali works and Playfair thought that should be the first object for legislation. Two other distinguished scientists, Hofmann and Frankland, confirmed Playfair's opinions.[6] So, this time—and unlike the position in 1846—the scientific advice was unequivocal. A further favourable testimony was that the alkali traders themselves wanted compulsory abatement of their acid fumes, if only to put an end to troublesome prosecutions by aggrieved landlords.

This and other evidence emboldened the select committee to make unanimous recommendations which, though they seem modest enough in hindsight, were at the time radical; and they set a pattern which still prevails in Britain's policy for the environment. The Committee did not propose a comprehensive law to cover all noxious vapours from trades and manufactures. Instead, it adopted a down-to-earth piecemeal approach for serious emissions (let us tackle the nuisances, one at a time, that we know we can control). The three nuisances that seemed ready to be tackled were noxious gases from the manufacture of soda, sulphuric acid, and ammonia alum. Substantial penalties were prescribed for failure to control these nuisances but it was left to the industrialist himself to decide how to abate his nuisance; the Committee did not want parliament to start dictating specific technical cures.

As to the enforcement of its proposals, the Committee made a recommendation of great historical importance in this story, namely that the enforcement officers should be 'wholly independent of all local control, and removed, as far as possible, from all local influence'. It was not, the Committee was at pains to say, that local authorities were not to be trusted; it was that officers paid by local authorities could not, by their very position, supervise enforcement, have rights of access to the factory run by the local owner (who might well be chairman of the local courts), and be expected to act with detachment. Nor could most local authorities expect to recruit highly qualified chemists as easily as they could be recruited by a department of central government. So although the suppression of general nuisances was to remain a duty for local government, the suppression of the 'monster nuisance' was to be the responsibility of some unspecified central authority. There was a further advantage in having inspectors attached centrally. The 'monster nuisance' was not confined to one corner of Britain: its pestilence covered all areas where alkali plants were multiplying, and (as one of the witnesses to Derby's committee said): 'It is remarkable how little is known in one place of what has been done in the way of suppressing nuisances in another place, by men carrying on a similar manufacture.'[7] A central inspectorate could ensure consistency, exchange information on control techniques, and acquire an expertise that would not be possible among officials acting for a score or more local boards of health, boards of guardians, and the like.

The report of Lord Derby's committee moved the government to act (Palmerston was now Prime Minister and would obviously have been sympathetic toward practicable legislation). On 23 March 1863 Lord Stanley of Alderley, as spokesman for the government, introduced into the House of Lords the Alkali Works Regulation Bill. It fell short of the committee's recommendations, for it covered only the emission of hydrochloric acid from alkali works (postponing any control of the other two emissions, those from the manufacture of sulphuric acid and of ammonia alum) and it did not concede the right of any person to take offenders to court. But in essentials, the government Bill dealt with the 'monster nuisance'. It laid down an emission standard—one, apparently, already agreed beforehand by the trade—namely that 95 per cent of the hydrochloric acid should be removed from the gases escaping from the chimneys of alkali works; and to enforce the law, the Bill provided for the creation of an inspectorate under the Board of Trade, with responsibility for all alkali works in the Kingdom. Of course the Bill was criticized in debate; landowners in the House of Lords complained that it was too timid; industrialists in the House of Commons complained that the choice of tribunal to try cases (the Court of Quarter Sessions) would bias judgements in favour of landowners. The Bill was amended to meet this latter complaint: the tribunal was to be the County Court, with unlimited rights of appeal to higher courts.

On 28 July 1863 the Bill received the royal assent, as the Alkali Act, 1863.[8] The Act made explicit a social attitude that had been implicit for some years, namely that central government ought to do something about the protection of air against pollution by noxious vapours. Members of parliament realized that they were embarking into uncharted waters. For the first time inspectors were empowered to enter factories, not on behalf of workmen (as had been the practice since 1833 under the Factory Act) but on behalf of something inanimate—the atmosphere and property damaged by acid fumes. And the inspectors were to be experts serving under a central government department, not under local authorities. So it was to be expected that the new law would be regarded with some apprehension, as an experiment, and it was ordered that it should continue in force from the time it came into operation (1 January 1864) until 1 July 1868, 'and no longer'.[9] In the event it led to a style for the politics of pollution-control in Britain which we have followed ever since.

A pioneer scientific civil service

Only a few records of the origin of the Alkali Inspectorate have been preserved, but we can piece together something of the story from what remains. First, the Board of Trade had to get Treasury approval to cover the salaries of the Inspector and his staff. 'Bearing in mind the extensive scientific knowledge combined with high character and great judgement required' for an inspector, the Board, basing its figure on salaries paid to inspectors under the Factory Acts, proposed £1000 a year for the Inspector himself, and £500 a year for sub-inspectors, together with travelling expenses.[10] The Treasury was unimpressed by this comparison and proposed salary scales of £700 and £350 respectively. The Board acquiesced reluctantly in £700 for the inspector but held out for £400 for his assistants, with a warning in its reply to the Treasury which shows that it was aware of the novelty and delicacy of this operation: 'It should be remembered', the Board's minute ran, 'that the Bill of last Session is one of an exceptional character and could not have been carried through Parliament without the cordial cooperation of the Alkali Trade who attach great importance to the scientific knowledge possessed by the Gentlemen who are appointed.'[11] The Treasury demurred also about the number of sub-inspectors the Board asked for, but again the Board held firm and four men were appointed to these posts.

The first Inspector, Dr Robert Angus Smith, was already well known as a sanitary chemist. The comment of the *Chemical News* on three of the other men appointed was that they were 'entirely unknown to us'.[12] All of them stayed to make their careers in the Inspectorate and one of them, Fletcher, succeeded Angus Smith and developed, with even greater skill, the style which Smith created. It was Smith who moulded the Alkali Inspectorate into the shape it still possesses today. The office of Inspector was, as one of Smith's admirers wrote after his death, one 'which was perhaps created for him to fill, and for the proper filling of which it may well be said that he was specially created'.[13] So what kind of a man was Smith?

From his father he inherited a Celtic fervour combined with a puritanical integrity; conscientiousness to a fault (parts of his annual reports on the Inspectorate are tedious and flat because he could not bring himself to discard irrelevant data); and a splendid physical constitution (it was said of Smith's father that he could walk 30 miles at a stretch when upwards of 70 years old, and Robert had similarly 'remarkable peregrinating powers').[14] From his mother he inherited charm, benevolence and toleration, the gift of patience, and such unruffled temper as to have earned for him the nickname 'Agnes' Smith. Born near Glasgow in 1817, the twelfth child of a severely Calvinistic millowner, he was sent at the age of twelve—as Scottish children were in those days—to the local University, with the intention, apparently, of becoming a minister. He abandoned the university after one year of a course in classics. Thereafter he took jobs as a tutor, and though it is said that he still contemplated becoming a minister as late as 1842, one of his tutoring jobs changed the course of his life. In 1839 the family to which he was attached as tutor (the Revd H.E. Bridgeman) took him to Germany. He decided to remain there for a time, to study chemistry under Liebig in

Giessen. Liebig was one of the founders of organic chemistry and at that time his laboratory was a place of pilgrimage for chemists from all over Europe. Not only was he among the illustrious pure scientists of his age; he was also one of the first chemists to apply his science to agriculture. So Smith found in the laboratory at Giessen an inspiration to apply science to social needs. More than that, he met contemporaries there who had an important influence on his career. Among them were Lyon Playfair (whose own career was determined by his stay in Liebig's laboratory), August Wilhelm von Hofmann (who came to England to found the Royal College of Chemistry in 1845), and Edward Schunck, who was to become Smith's lifelong friend in Manchester. In 1841 he was awarded a Ph.D. in Giessen and in 1843 he went to Manchester (where he settled for the rest of his life) to become assistant to Playfair, who by then had been appointed honorary professor of chemistry at the Manchester Royal Institution. It was a happy moment for this reunion with Playfair, for in the summer of 1843 Playfair was appointed to the commission on the Health of Towns. Angus Smith helped him in this work; this was his apprenticeship in sanitary science and it settled the shape of his career.

For some twenty years, from the mid 1840s to the mid 1860s, Smith made his living as a consultant chemist. Many of the problems brought to him concerned public health: the contamination of water supplies, the management of sewage, the pollution of air. In common with his contemporaries, he believed that disease was spread by 'miasma'; a bad smell was the warning of disease. He became interested, therefore, in the chemical analysis of air and water. His grasp of chemistry was unsophisticated and soon became outdated, but he had a flair for the application of simple chemical principles to practical problems; sagacity and sound common sense were his outstanding virtues. He was a prolific writer (he published over 60 scientific papers, some of them of unendurable prolixity); but his tact and reasonableness in dealing with people and his kindness and encouragement to the young, endeared him to all who got to know him.

These qualities, rare as they are, were not sufficient to qualify Smith for a professorship (his style of lecturing was rambling and diffuse and his research work was undistinguished), though he did apply twice for the chair of chemistry at Owens College, Manchester, and was, quite rightly, passed over in favour of Frankland on the first occasion and Roscoe on the second. But neither Frankland nor Roscoe could have done the job to which Smith was called in 1864; for this job Smith had precisely the required qualities.

It is not difficult to imagine the obstacles Smith had to overcome. An isolated government official based in Manchester, with very little backing or guidance from his employers in Whitehall, 180 miles away; empowered to control emissions from a great and flourishing industry. His only hope was to secure the confidence and co-operation of the factory owners. One tactless letter, one injudicious prosecution for infringement of the Alkali Act—and Smith would have had the whole alkali industry ganged up against him. Yet to be lax and indulgent in the enforcement of the law was out of the question for a man like Smith. He had no precedent for what he had to do; 'no one knew what my duties could be', he wrote toward the end of his career, 'I had to teach myself as well as others'.[15]

The record of Smith's definition of his duties is to be found in his annual reports.[16] They trace the evolution of a national policy to reconcile the exploitation of the environment for the benefit of industry with the protection of the environment for the benefit of society; an evolution that was guided for twenty years by one man, with hindrance rather than help from parliament. The daily routine was onerous and often uncomfortable. Between March and December 1864 about a thousand visits were made to the 84 alkali works registered under the Act. Smith himself visited each works at least once. Some visits were made on Sundays and some at night. The work involved perpetual travelling, made worse by the load of equipment the inspectors had to carry, 'especially when a visit is to be paid to works not easily accessible and near railway stations where no cabs are to be found...'[17] And the testing itself could be a pretty unpleasant process. On Tyneside it was common for condensers to have an opening at the top for the escape of uncondensed gases, and this is where the samples had to be taken. An objection to this plan, wrote Smith in one of his reports, 'is the difficulty of mounting to the summit, and when there of working calmly at a height of 125 feet on a platform slenderly railed under a strong wind and even rain. One may occasionally stand for an hour under these conditions...and as a rule it may be said that inspectors who are not equal to sailors in climbing cannot make examinations at the summit of the towers. Even sailors could not take up the apparatus. This, however, is only an inspectors' difficulty which must be overcome.'[18] Notwithstanding these hardships, Smith was meticulous in his insistence on care and reliability in sampling. Here are a couple of examples, again from Smith's annual reports:

> Circumstances compelled Mr. Todd [one of the sub-inspectors] to give up the new aspirator; indeed he broke a blood vessel by the extra exertion required in shaking it to absorb the vapours: but in making a change he did not return entirely to his previous method, and I have not the same confidence in all the results.[19]

> If the condensation is not perfect, although as complete as the Act requires...the examination often made in mud and rain is a matter requiring a good deal of patience...[20]

These visitations taxed not only the physical endurance of Smith (already 46 years old when he was appointed) and his staff; they were a strain on the temperament, too, for it was essential not to alienate the managers of the works. 'Beginning as I did', Smith wrote in 1864, 'with a strong desire to aid the public in the way most agreeable to the manufacturer, I could not forget that it was in his power...to make the office of inspector most difficult and disagreeable.'[21] Wisely, he decided to soft-pedal on prosecutions, and relied on what he called 'advice and friendly admonition'.

The strategy worked. By the end of the first year Smith was able to report that the escape of hydrochloric acid, previously estimated at some 1000 tons a week, had been reduced to some 43 tons. Every alkali work had condensed 95 per cent or more of its acid emissions. Unfortunately this was not accompanied by any dramatic improvement in the state of the atmosphere. There were two reasons for this. First, new works were being opened, new chimneys were going up to eclipse the benefits of what Smith described as the 'rather violent exertions' of his first year in office. Second—and much more important—were exasperating constraints in the Act itself. All sorts of other noxious vapours were polluting the air and Smith was powerless to do anything about them. His jurisdiction was solely over hydrochloric acid, and, even for that acid, only over emissions from alkali works. This led to vexatious incidents. Thus, an action was brought by a Mr Powell against Messrs Chance for damage from acid vapours. To Smith's dismay he found no escape of hydrochloric acid. The nuisance was due to sulphuric acid, which was not covered by the Act. So Mr Powell lost his case and Smith, anxious to keep within the law, wrote:

> It is not for myself and colleagues to become general accusers, and we do not think it proper to point out the offence even when it is known to us, unless we find muriatic acid.[22]

But in fact he was already finding the law irksome. Even the quantified emission standard had its drawbacks. True, it protected the manufacturer from mischievous litigation but, as Smith went on to say:

> A strange result, therefore, takes place; we become defenders of the alkali works, and appear also to defend the right of sending out 5 per cent of the gas; whereas the case is merely this: we cannot publicly object until that amount is attained.

Here was the perpetual dilemma for the alkali inspectors. On one side they were peppered by complaints from the public that noxious vapours were not being properly controlled (with the innuendo that the inspectors were not doing their job): on the other side they were hobbled by the constraints of the Act, which, throughout Smith's tenure of office, parliament only rarely and grudgingly loosened. Smith never took much interest in politics and lacked the guile that might have won for his Inspectorate some concessions in Whitehall. But even after one year in office he had decided it was his duty to interpret the spirit of the Act rather than to obey the letter; and the spirit lifted him far beyond muriatic acid. Two years later he was beginning to see his tiny experimental Inspectorate as the embryo of a comprehensive system of control which would ultimately embrace all pollution of air and water. His inspectors were already exceeding their brief by making unofficial inspections of noxious vapours other than hydrochloric acid. Why should there not be fixed emission limits for all these vapours? In the report for 1866 he wrote:

If a similar fixed point could be adopted in the case of every gas, there would be complete protection to the public and manufacturer on both sides up to that point...It seems to me a most important thing to seek such fixed points, and where they cannot be attained to make the nearest approach to them...These fixed points may be sought both for air and water. I have attempted a beginning, and will develop it if it promises well.[23]

By 1868 he was including some consideration of the composition of coal smoke, though he felt obliged to justify this: 'It may be said not to lie in my province, but the escape of muriatic acid into it occasionally was the cause of my attention.'[24] Every annual report carried hints that his powers should be enlarged:

I have thus extended my work from that of the inspection of alkali works to the examination of works generally for sanitary purposes, so far as chemical means are concerned. It may then be asked, why do I not proceed with the inquiries, if I consider them so important? This is easily answered. The work has not become official, and individual work must be intermittent.[25]

In this probationary period the office of Inspector was not intended to be full time (though Smith made it so) and this enabled Smith to declare that he was doing unofficial 'individual' work beyond his statutory powers.[26] But all along he made no secret of his intention to go on pressing for this unofficial work to be brought within the scope of the Act. In his report for 1879 he wrote: 'as inspector, my duty has always been supposed by myself to include a study of air and climate in connection with manufactures...' As he and his junior colleagues lugged heavy equipment from railway stations to factories, clambered up towers in all seasons to fill aspirators with exhaust gases, carried their samples to makeshift laboratories (often in the inspectors' own homes) for analysis, Smith never lost sight of his vision of an Inspectorate which would have power to control all noxious vapours discharged into the atmosphere from industry. History has vindicated his aims, though they have had to fight their way to success through many obstacles.

This vision was one characteristic of Smith's administration of the Act. Another characteristic was his style of negotiation with the manufacturers. After a decade of working, Smith wrote:

There are two modes of inspection, one is by a suspicious opponent, desirous of finding evil, and ready to make the most of it. The other is that of a friendly adviser, who treats those whom he visits as gentlemen desirous of doing right...The character of the inspection which I have instituted is one caused partly by my own inclination, and partly by the nature of the circumstances.[27]

This trustful approach, combined with the fact that the expansion of industry cancelled out some of the benefits of the inspectors' efforts, laid the inspectors open to criticism that they were in the pockets of the manufacturers. Prosecutions were rare (there were only nine in the whole twenty years of Smith's service). This, in Smith's view, was a measure of success: a multitude of prosecutions would have been a sign that the inspectors were failing, not succeeding, in their task of educating manufacturers. The public, Smith wrote, 'are continually confusing or attempting to confound our duties with those of the police...The Government inspectors have been more as teachers raising up the standard of labour in the works'.[28]

The experiment is endorsed

It was with some apprehension that the alkali inspectors approached the year 1868, for the Act was due to expire after 1 July of that year. 'I cannot shut my eyes', wrote Smith in his first annual report, 'to the fact that inspection of this class is upon its trial before the country'.[29] However, Smith's annual reports had been well received in parliament. Both Lord Ravensworth (whose property lay in a region suffering from alkali works) and Lord Derby spoke approvingly of the Inspectorate's work.[30] Noxious gases were still damaging vegetation but it was recognized that this was not the fault of the Inspectorate. So there was no opposition to a simple Alkali Act Continuance Bill introduced into the House of Lords on 14 May 1868. Indeed the title of the Bill was changed from 'Continuance' to 'Perpetuation' and in that form it received the royal assent on 25 July.[31] Some speakers in the debate wanted the Bill to be expanded to cover processes other than soda manufacture, but the government was chary of any change which might raise unforeseen technical difficulties. The sole dissent came from the member for Finsbury, who questioned whether it was proper that the cost of the Inspectorate should be borne by the public rather than by the beneficiaries, and hoped (in vain) 'that a clause would be introduced to remedy that state of things'.[32]

So ended the novitiate of the alkali inspectors. Their duty was still a very narrow one but they had received encouragement, for when the 'Perpetuation' Bill was debated, the government announced its intention to ask for a report to be made to the Board of Trade on other gases produced by manufacturers.[33] Angus Smith promptly accepted this invitation; but of course it was not possible to study other gases unless inspectors could have entry to other sorts of works. So he renewed his plea for wider powers for the inspectors to be able to enter and examine without bringing any action or appointing any penalty.[34]

The Public Health Act of 1872 lifted the Inspectorate from the Board of Trade and put it under the Local Government Board. This stimulated Smith to step up pressure to have the Alkali Act amended. He said frankly that the Act as it stood was unfitted to deal with the great increase in the chemical industry: 'there are districts in which more injury has been done to vegetation this year than on any previous. I refer chiefly to the neighbourhood of Widnes. That town is a battery of works, and sends vapour from many chimneys, and constantly.' One weakness in the Act was that it reduced the noxious vapours only by a percentage of the total emission. Smith proposed a more stringent standard for hydrochloric acid, namely 0.2 grains of acid per cubic foot (a figure already acceptable to the trade). For other vapours, about which far less was known, he proposed 'treatment by the best methods' (leaving his options wide open) combined with authority for himself and his staff to inspect the works.[35] Meanwhile trade in soda and other chemicals was booming, new works were put up with insufficient capital and no adequate control against pollution, and the public began to clamour for better control by the Local Government Board. A pressure group, the Alkali Act Extension Association, was set up for Northumberland and Durham, under the chairmanship of the Duke of Northumberland. This was, of course, a welcome reinforcement for Smith's campaign. But about this time the whole matter was thrown into some confusion by a major reappraisal of the machinery of local government. John Simon, one of the great figures in sanitary reform in the nineteenth century, had been brought into the Local Government Board as secretary of the medical department some years earlier and had reorganized local health authorities so as to give them more authority and more responsibility. It was part of this reorganization which brought the Alkali Inspectorate (and water authorities also) under the Local Government Board. The next logical step, in Simon's view, was to delegate to local health (or sanitary) authorities the responsibility for controlling all chemical nuisances and he did not see why the control of vapours from alkali works should be any exception. Simon put forward his views with characteristic force and clarity. The task of central government, he wrote, was to inspect nuisance authorities, not nuisances themselves; if it went beyond this duty, it hindered local authorities from acquiring a proper sense of duty to the public. The argument that local authorities might find it difficult to discharge the duties delegated to them because of strong local interests in support of manufacturers was no stronger for vapours from alkali works than it was for other nuisances (foul water, for instance) which the local authorities were expected to suppress. The argument that officials attached to local authorities would lack sufficient knowledge of chemistry to control noxious vapours could be countered by enabling the local authorities to hire consultants. In Simon's own words:

The great aim of the central government has been, and, I think, should still be, to excite, and in case of need compel, the local authorities to act vigorously and impartially in the face of these several interests...[To fail to do this would] contribute to a permanent demoralization of local authorities.[36]

Simon's policy for devolution of powers to local authorities was the received doctrine of the Local Government Board. It was the antithesis of the policy forcefully propagated by Edwin Chadwick, who had a decisive influence on the decision, under the Factory Act of 1833, to appoint four factory inspectors attached to the Home Office, 'the first attempt', wrote J.R. Green, 'to establish a new and vital principle of central control'. For a moment it must have seemed as though the vision of Angus Smith was going to dissolve. But Simon's view did not prevail; his influence at the Local Government Board was already on the wane. To have liquidated the Alkali Inspectorate, just as it was gaining the confidence of industry, would not have been a wise move politically. Besides, Smith and his colleagues had accumulated a unique experience of the problems of industrial pollution of the air, and Smith had demonstrated (in his report for 1867) how much more effective the Alkali Act was, compared with analogous legislation in other countries. He quoted as an example what had happened (or, rather, not happened) in France.[37] In 1810 Napoleon issued a decree: 'manufactures which give out an unwholesome or unpleasant odour will not be allowed to be established without the authority of the Administration'. Manufactures were to be classified in three categories: those which ought to be distant from all private dwellings, those which ought not to adjoin private dwellings unless conducted so as to cause no nuisance, and those which might remain without inconvenience near private dwellings. There followed a list, largely expanded by 1867, of the three categories of manufactures. It was a typically elegant Gallic law but, as Angus Smith concluded after examining its impact: 'comparing the attempts to legislate with the result, there has been... a failure of the most decided kind. The so called preventive system has failed.' By trial and error Angus Smith was discovering the virtues of the British penchant for pragmatism.

3

THE BEST PRACTICABLE MEANS

Growing discontent

The Act to perpetuate the Alkali Inspectorate left untouched the narrow scope of the original Act of 1863. Year after year, in his annual reports, Angus Smith pressed for a widening of this scope. Meanwhile complaints to the Local Government Board increased and were getting unwelcome publicity. There were ominous signs that pressure groups were being organized. So, once the decision was taken not to throw the Alkali Inspectorate into the stream of local government, the Local Government Board set about drafting a Bill to extend the original Act. It was a feeble draft;[1] clearly the Board had decided to keep to a minimum its interference with the freedom of manufacturers. The new clauses represented the very least the state would be obliged to do to placate public criticism. There were only three significant innovations; one prescribed an additional emission standard for hydrochloric acid from alkali works that Smith had recommended and industry had already tacitly accepted, namely the standard (still current) of 0.2 grains per cubic foot; a second innovation was to add one more process to come under inspection, namely the production of copper by the 'wet' process (which produced hydrochloric acid that could be controlled just as it was controlled in alkali works); the third innovation seemed a mere mouse of a measure, for it applied only to alkali works—noxious vapours from all other manufactures were left uncontrolled—but it was a clause of great significance for the future of the Inspectorate. It ran as follows:

> the owner of every alkali work shall use the best practicable means, within a reasonable cost, of preventing the discharge into the atmosphere of all other noxious gases arising from such work, or of rendering such gases harmless when discharged.

The noxious gases for this purpose were sulphuric acid, sulphurous acid (unless it came from the combustion of coal), nitric acid or other noxious oxides of nitrogen, sulphuretted hydrogen, and chlorine.

Smith must have been disappointed. The Bill added only one new process to the jurisdiction of the inspectors, and, for other gases from alkali works, the Bill put a new responsibility on the shoulders of the inspectors while at the same time denying them the authority to match the responsibility; for the phrase 'best practicable means' had, ever since the Leeds Improvement Act of 1842, been a secure defence for polluters brought to court for creating smoke.* The wretched phrase even weakened the chances of a

*Except in London, where the strict administration of Palmerston's Act did produce a gratifying crop of convictions.

conviction under the common law of nuisance. It was the greatest contribution of Smith and his successors to the protection of Britain's atmosphere that this phrase, under their administration, was changed from being a shield for polluters to become a weapon for those who controlled pollution.

The Bill encountered little serious opposition. The phrase 'within a reasonable cost' was deleted in the House of Commons (it was otiose anyway, for 'practicable' implies that the cost is reasonable). In the House of Lords the Bill was criticized as inadequate to cope with the ever rising scale of chemical industry.[2] Lord Ravensworth moved to add a clause which would give any person suffering from noxious vapours from alkali works the right to complain to the Local Government Board, which in turn would direct an inspector to report upon the complaint.[3] This was supported by several speakers—a sign of growing discontent with the administration of the Act—but it was in the end rejected because it implied that the inspectors were not doing their duty and it would have been objectionable to embody this innuendo in an act of parliament.

The Alkali Act, 1874, came into force on 1 March 1875.[4] Five months later the great Public Health Act which consolidated the sanitary legislation of England and Wales was put into the statute book. This fixed for the moment the dichotomy of control of air pollution: on one hand special emissions from two industrial processes were controlled from Whitehall; on the other hand all other pollutants—smoke, noxious gases, odours—were controlled locally.

Before Smith had had a chance to make use of the slight—very slight—easement of constraints which hamstrung his inspectors, complaints began and were passed on to the Inspectorate. Smith, smarting with grievance, defended himself:

> I suppose I must forgive the ungracious character of the enquiry...it may not have occurred to any one that an Act of a progressive character like that of 1874 cannot be brought to perfection on an appointed day...On the first of March we had power to begin and we began...already furnaces have been stopt, improvements at great expense have been made...[5]

He repeated his conviction that all chemical works giving out offensive gases 'should, on complaint being made, be put for a certain time under inspection. This prevents the demand, at least for the time, of a great number of inspectors.'[6] This last sentence was a capitulation to the niggardly attitude of parliament and the civil service toward any policy for clean air.

There followed a crescendo of complaint to the Local Government Board. On 17 February 1876 the President of the Board, Sclater-Booth, was challenged in the House of Commons to bring in better laws to control noxious vapours. A week later Lord Winmarleigh presented a petition in the House of Lords from a pressure group in Lancashire and Cheshire, praying for an amendment in the law, and he urged the government either to act or to appoint a royal commission.[7] On one estate in Cheshire, he said, trees were 'like a forest of masts even in midsummer'. Thereupon the Duke of Northumberland gave notice of his intention to move for a royal commission.[8] In March another pressure group was created in Liverpool; following this a large deputation of landowners and local authorities from Northumberland, Lancashire, Cheshire, and elsewhere, led by Earl Percy, crowded into the rooms of the Local Government Board to press for action straight away. Lord Winmarleigh, who accompanied the deputation, again urged the need for a royal commission. Sclater-Booth was evasive: 'he did not see any urgent necessity for a Commission as he thought a great deal could be done with the information already in hand'. He said he had already ordered the preparation of a small Bill to deal with some aspects of the nuisance (the most scandalous one was the emissions of sulphuretted hydrogen—a nasty gas smelling like rotten eggs—from factories especially in the neighbourhood of Widnes) and that he would 'take that up again' but he dampened any expectations by a parting shot: he 'did not think the House of Commons would look with satisfaction on any proposal to increase the number of inspectors'.[9]

The noble victims of noxious vapours were not going to be put off by inertia in the Local Government Board. Ten days after Earl Percy's deputation had put its case and received what amounted to a brush-off from the President of the Local Government Board, the Duke of Northumberland rose in the House of Lords to propose a royal commission.[10] His overture was to present a petition, similar to the one presented from the north-west of England, on behalf of the north-east: 'the evil complained of is widely spread, and whether on the East or West side of the island, its effects are found intolerable by all classes'. His original intention had been to ask for a commission to look only into nuisances from mineral works, but he enlarged the proposed terms of reference to cover 'all such emanations from whatever trade or works arising'. In the last census the number of employees in mineral works had increased by 84 per cent and in alkali works by 35 per cent; and the legal remedies, whether under common law or statute, were 'practically almost inoperative' to bring redress to those whose property or health suffered from pollution of the air by noxious vapours from industry. The Duke was supported by other members of the House. The government was evidently resigned to acquiescence, for the Duke of Richmond promptly announced its intention to agree to the motion for a royal commission. The motion was agreed to and the Commission duly appeared—a trifle tardily—four months later, on 18 July 1876.

The Alkali Inspectorate after ten years

For just over a decade the Alkali Inspectorate had been on trial, under the first Alkali Act of 1863, confirmed by the Perpetuation Act of 1868, and marginally extended by the amending legislation of 1874. Angus Smith had patiently moulded its style of working and unswervingly declared its ultimate purpose. The Inspectorate was now to face a searching reappraisal. It was still the same tiny band of men who had been appointed in 1864, though the number of registered works under inspection had increased from 84 to 160 and the amount of capital invested in the alkali trade had trebled, from about £2 millions to about £6 millions. The achievement of Smith and his colleagues was hidden by this great expansion of the industry; noxious gases were still escaping; trees and crops were still withering under the acid vapour. The only visible evidence of achievement was in the pages of Smith's annual reports. Verbose, rambling, repetitive though these reports are, they do impress anyone who plods through them with his achievements, difficulties, disappointments, and frustrations.

The achievements of the work were these: first, to have won the confidence of the trade; second to have persuaded the alkali makers to set up laboratories and to employ chemists to take daily measurements and to keep an 'escape book' as regularly as a ledger; assisting science, as Smith put it, to enter the industry; third, to have convinced manufacturers that it was in their own interests to recover the acid vapours and to improve on the devices for doing this. Finally the inspectors fulfilled their prime duty: despite the inevitable number of exceptions, gases escaping from works registered for their inspection really did not contain more than 0.2 grains of acid per cubic foot; and this result was achieved through consent, not prosecutions. People who took the trouble to inform themselves about the matter knew this and acknowledged it. The Duke of Northumberland, for instance, moving for a royal commission, went out of his way to say that his motion was in no sense a criticism of the inspectors. They have, he said, 'discharged their duty very fairly on the whole; they are greatly overworked, and yet have greatly diminished the cause of complaint...they wish to lead rather than drive men, which I believe, where great interests are involved, to be the best and wisest way of proceeding'.[11] Smith's peers had a similar opinion. In *Chemical News* there was a eulogy of Smith in 1875:

> If the Act is successful the result is mainly due to his zeal, tact, and intelligence... He does not seek to lay down at once a hard and fast line, but as a truly practical man he aims at and effects gradual improvement. We can wish nothing better for the cause of sanitary reform than that all inspectors who have to deal with 'standards' may sit at his feet.[12]

We have already said something about the difficulties and dangers of the work. The perpetual disappointment about the work was the crippling constraint of the Act, putting out of reach of the inspectors all sorts of polluting industries and—until 1874—all gases except hydrochloric acid, even in the alkali industry.

As for the frustrations, surely the chief of these was indifference amounting to negligence in Whitehall, manifested most vividly in a persistent refusal to allow Smith to recruit more assistants or to pay adequately the sub-inspectors he already had. Despite the modest salaries fixed by the Treasury at the beginning, ranking them below factory inspectors, and despite repeated appeals for increments in salary, none were forthcoming until 1876, and then only for the sub-inspectors. On two occasions the sub-inspectors approached the Board of Trade for an increase in salary.[13] In 1866 they were told they must await the revision and probable extension of the Act—an extension which never materialized. In 1871 their appeal was again rejected. After the passage of the Perpetuation Act in 1868, Smith requested the Board to appoint him on a full-time basis (he had of course been working full time since he was appointed but the post was deemed not to be a full time one in the experimental stage), and to increase his own salary.[14] This also was turned down. The culminating snub came in 1875 when Fletcher (who subsequently succeeded Smith) made a private approach to the new masters of the Inspectorate, the Local Government Board, for an increase in salary after 11 years' service and after the Act had been amended to augment the inspectors' responsibilities.[15] The Permanent Secretary, Lambert, passed Fletcher's note to the President with an unhelpful comment, and the President (Sclater-Booth) added this minute:

When it appears evident that the labours of these Inspectors are producing an appreciable result, I shall be happy to take the question of increasing their salaries into considn.[16]

Even the patient Angus Smith was moved to protest at this exasperating insensitivity toward his staff.[17] He warned his masters that he would not be able to keep his team together if they were exposed to this sort of misunderstanding. His retort, too gently worded to compel attention, cut no ice with Sclater-Booth, who sent this minute to his Permanent Secretary:

Say that the complaints made to me by deputations last year were to the effect that the ravages caused by the fumes of muriatic acid etc from the alkali works were doing most serious injury.

This was particularly the case on the Mersey which I suppose is within Mr. Fletcher's district.

If I hear that the vegetation of the coming Spring is perceptibly the better for the exertions of the Inspectors under the provisions of the new Act, it will form good ground for applying to the Treasury in the matter of these salaries.[18]

But this was not an argument to appeal to Angus Smith: 'It was not our duty to cause crops to grow, it was our duty to cause the works under our charge to send out not more than 5% of muriatic acid.' The sub-inspectors' salaries, but not the salary of Smith himself, were ultimately increased (in March 1876) by £100 a year,[19] but they were still less than the salaries paid to inspectors of factories; for the sub-inspectors £100 lower and for the Inspector £300 lower. What evidently pained Smith even more than the inequity about salaries was the Local Government Board's churlish indifference to the significance of his work. In 1883, after nearly twenty years' service in the Inspectorate, Smith accompanied a proposal to the Board with this moving lament:

I should be very much pleased if you favoured me with your opinions on any of these points, one likes at times an outside view. I am always at work well or ill, at home or not, and I attend only to my view of the subjects, with Chemical aids of course, knowing very little of gossip or what I am 'as others see me'.[20]

A royal commission

The Royal Commission which began its work in 1876 did provide the kind of 'outside view' that Smith and his colleagues needed.

The terms of reference for the Royal Commission were based on the Duke of Northumberland's motion in the House of Lords:

to inquire into the working and management of works and manufactories from which sulphurous acid, sulphuretted hydrogen, and ammoniacal or other vapours and gases are given off, to ascertain the effect produced thereby on animal and vegetable life, and to report on the means to be adopted for the prevention of injury thereto arising from the exhalations of such acids, vapours and gases.

Hydrochloric acid was not specifically included in this list of noxious vapours, perhaps because the Local Government Board wanted the Commission to concentrate its attention on processes not covered by the Alkali Act. But the commissioners, very properly, began by reviewing the existing system of control and proceeded from there to less familiar problems.

It was a well balanced body of commissioners. Lord Aberdare, the chairman, had at one time been a liberal MP for Merthyr Tydvil, an industrial area in Wales. He had intellectual interests (he was made an FRS in 1876, and became the first chancellor of the University of Wales in 1894); he had taken part in the debate that set up the Commission; and he took the responsibility seriously 'at some sacrifice', he later confessed, 'of personal comfort and almost of health'.[21] To balance a Liberal peer there was a Conservative, Viscount Midleton. There were three members of the House of Commons, all from constituencies affected by industrial pollution: Earl Percy (son of the Duke of Northumberland); the Hon. W. Egerton, son of Lord Egerton who had been a member of Lord Derby's committee in 1862; and J.C. Stevenson, MP for South Shields. Together with these five members of parliament there was a retired Admiral of the Fleet, W.W. Hornby, and there were three scientists: F.A. Abel, government chemist to the war department, and inventor of cordite; A.W. Williamson, professor of chemistry and practical chemistry at University College, London; and H.E. Roscoe, professor of chemistry at Owens College, Manchester.

The commissioners worked for two years and produced a weighty and thorough report. They were confronted by two questions that still trouble those who have to legislate to protect the environment. First, there was no indisputable evidence that noxious vapours (other than smoke and sulphurous gases) harmed human health. This being so, how far should parliament go to legislate for the preservation of private property or the protection of Nature? (Lord Aberdare had himself asked this question during the House of Lords debate.) Second, there were some flagrant nuisances for which there was at that time no technology for abatement. The only effective way to stop them would be to shut down the works. Should these nuisances be tolerated or suppressed?

The vapours may not have been a hazard to health, but they certainly were a hazard to property. The Archbishop of Canterbury caused a stir by complaining publicly about deterioration of the stonework at Lambeth Palace, and the discoloration of his knives and forks.[22] 'There is not a figure that has got a nose on it', declared the rector of Lambeth, 'or, if it has, at a touch it will drop off.'[23] The culprit for producing what was called 'Lambeth fog' was believed to be Doulton's potteries. Henry Doulton appeared personally before the Commission. 'Our operations are quite innocuous', he said, and he warned that it would be a serious matter 'to interfere with a trade which gives employment to a thousand families, and which has existed for centuries'.[24]

This is the ethical dilemma which politicians have to resolve when they legislate for the environment. Environmental problems do not pose a conflict between good and bad; they pose a conflict between one kind of good (unpolluted air for the Archbishop of Canterbury) and another kind of good (the trade in pottery and the employment of thousands of workmen). The point of reconciliation for this sort of conflict shifts as public opinion evolves. 'I suppose', asked Lord Aberdare at one hearing of the Commission, 'it is the fact that people, as civilisation advances, get more sensitive to evil smells?'[25] Exactly so; this was the political problem the Commission had to consider: what price should be paid for amenity?

The Commission's inquiries can be summarized under two headings: (1) a critique of the existing system of control of noxious vapours and (2) the question of its extension and change in its administration. Should it cover other works and trades? And should it be administered from the centre under a department of State or devolved to local authorities?

First, the existing system of control. The Commission heard many complaints about the persistence of noxious gases despite the work of the alkali inspectors, but it was evident that much, though not all, of this vexatious nuisance was due to pollution over which the Inspectorate had no control:

sulphuretted hydrogen, sulphur dioxide from the burning of coal, and the like. The commissioners' conclusion about the alkali inspectors was: 'we have no hesitation in expressing our opinion that their operation has been beneficial; and that even where their efforts have been counteracted by a great increase of works, they have been effectual in preventing a great amount of injury which would otherwise have been experienced'.[26]

There was one qualification to this favourable verdict, namely some question as to whether the Act would have been more effective if it had been administered more strictly. Angus Smith had often defended his policy of 'advice and friendly admonition' rather than prosecution, but some witnesses before the Commission were unconvinced; they thought the velvet glove strategy sometimes concealed weakness, and even if it had been desirable at the outset, it had, in the opinion of these witnesses, 'been unnecessarily prolonged'.[27] This is a charge still brought against the alkali inspectors. The commissioners gently questioned whether a stricter inspection would not have secured greater benefits,[28] but they were in no doubt about the danger of relying on prosecutions: zeal to prosecute would have created a spirit of opposition among manufacturers which could have alienated the trade against the whole idea of abating pollution.[29]

The commissioners' recommendations were a reassuring vindication of the policies of the Alkali Inspectorate. What the system needed was to be upheld and strengthened. To this end the Commission's report made four proposals. The first was to increase the number of inspectors. The second was to give them more powers over alkali works; these were to include the setting of emission standards not only for hydrochloric acid but also for sulphur (one grain per cubic foot) and for oxides of nitrogen (half a grain of nitrogen per cubic foot), and also power to inspect and report upon the nuisance caused by the waste heaps of alkali—a nauseous mess called 'galligu'—which were another by-product of soda manufacture.

The third proposal was an oblique way to embarrass polluters without going to the length of taking them to court; it was to publish all recorded escapes of noxious gases with the name of the works in which they occurred, and also to publish reports of defective machinery and its operation. This was a sound psychological proposal, for it was evident from some of the witnesses that the alkali traders themselves were keen on disciplining any of their own number who were infringing the law. The fourth proposal—the only one that was not unanimous—was made in the hope that it would encourage prosecutions of irresponsible offenders. It was to authorize local authorities, with the sanction of the Local Government Board, to take offenders to court under the Alkali Act (the Act permitted only the inspectors themselves to initiate legal proceedings.)

The second heading under which the commissioners assembled their conclusions covered the extension of the law to other works and trades and the pattern of administration of the service. That some extension was needed they had no doubt. But to make extension feasible two problems had to be solved. One was technical: how to fix emission limits for some of the noxious gases, and what those limits should be. The other problem was administrative: to decide what sort of authority should enforce the law. Ought it to be a central inspectorate attached to the Local Government Board? Or ought it to be devolved upon local government? And if to local government, how were the dozens of local authorities to be able to recruit competent chemists for the work of surveillance?

Let us pause to consider these two problems. If a fixed emission standard could not be set (owing to technical difficulties) was it good enough to use the formula 'best practicable means' which had been included in the amended Alkali Act of 1874? For smoke abatement the formula had rendered the law impotent. Angus Smith was instinctively critical of it. He preferred the less slippery variant 'best known means'[30] though he concurred in interpreting this as 'the best practicable method which might be known at the time',[31] and when pressing for an extension of control in 1875, disclosed his reluctance to resort to such phrases: 'the mode of introducing the other works, must, I suppose, be in a manner allied to the indefinite clauses of the Act of 1874'.[32]

Fletcher, Smith's assistant, had a much more optimistic view about the potential value of 'best practicable means'. Fletcher's evidence to the Commission was seminal, for he put forward an opinion which became—and still is—a linchpin in British policy for protecting the environment. His evidence ran:

> With regard to other vapours, called in the Act of 1874 'noxious gases', I think that the Act gives ample power for their control. In clause 5 it says that 'the owner of every alkali work shall use the best practicable means of preventing the discharge into the atmosphere of all other noxious gases, etc.' Some persons have expressed a fear that this is not sufficiently definite and binding on the manufacturer. For my part I feel it to be more binding than a definite figure, even if that could be given, for it is an elastic band, and may be kept always tight as knowledge of the methods of suppressing the evils complained of increases.[33]

For Fletcher, the formula 'best practicable means' could be successfully used as a permanent basis for the abatement of pollution, without prescribing fixed emission standards even when it was possible to do this. For Angus Smith, the formula was a temporary expedient, to be used until a firm basis for a fixed emission standard could be settled.[34] Fletcher's vision for the future of the Inspectorate was already better focused and more imaginative than Smith's. He would have liked a comprehensive Noxious Vapours Act, resting solely on the phrase 'best practicable means' (he believed it could be a weapon on the side of the Inspectorate in law suits) and 'running parallel with the Rivers Pollution Act of last session'.[35]

He even saw on the horizon the possibility of including coal-burning among the responsibilities of the Inspectorate, to be controlled by the best practicable means.[36] (This, as we shall discuss in later chapters, was to be a much more difficult business than the control of noxious vapours.)

What did manufacturers think about 'best practicable means'? Most of them were in favour of it. 'I think', said a witness from bleachers and chemical manufacturers in Manchester, 'that that Act which obliges all works to adopt the best practicable means of remedying the nuisance is an excellent form of legislation.'[37] There was, of course, the problem of how to define exactly what the best practicable means were. But another industrialist had the empirical answer to this problem: 'Supposing the inspector finds one [factory] working, say, at one grain [he was referring to sulphuric acid per cubic foot of gas] and another at five grains, I do not know whether he has power to compel the five grains worker to come down to the one grain worker; if he has such power, then I think that a maximum need not be named at present.'[38]

Not all witnesses were satisfied with reliance on a phrase so vague as best practicable means in a revised Act, but their arguments against it were weak. In the end it stood up well to this rigorous cross-examination, though the report is guarded about its adoption as the sole criterion. Applied to smoke it had, except in London, a poor record. Applied to noxious gases it had been on the statute book for less than two years: it was too soon to pronounce a final verdict on it. So the commissioners did not accept Fletcher's radical proposal that fixed emission standards should be abandoned in favour of best practicable means. Instead, they recommended a more cautious and traditional pattern, namely, that there should be a three-tier category of control of noxious vapours: fixed limits wherever they could confidently be prescribed, inspection and insistence on the use of best practicable means for some other emissions, and inspection alone for emissions where there seemed no prospect of applying any agreed best practicable means.

Over the pattern of administration, it was plausible to suppose that there would be a massive increase in the amount of inspection required if control were to be extended to all sorts of other processes. Should it be control from the centre, as it was for the Alkali Act? Or should control be a power delegated to local government?

The alkali inspectors themselves were willing, indeed anxious, to take on the extra responsibility. Smith told the Commission that the primary purpose of inspection was not to act as policeman but as teacher, to persuade and to help the alkali industry to control pollution; in so far as inspectors succeeded in this purpose, the frequency of visits could be reduced.[39] Both Smith and Fletcher believed that half a dozen more scientifically trained assistants would be enough.[40]

Manufacturers, landowners, and officials of local government were practically unanimous in their preference for the control of noxious vapours to be under the central government. This, they emphasized, would ensure a high level of expertise, insulation from local pressures, and a uniform interpretation of the act in different parts of the country. Major Cross, chairman of the local authority in Widnes (a centre for the alkali industry) put the case against devolution to authorities like his own very clearly. The sanitary authority in Widnes was about half composed of alkali manufacturers and:

It would never work. In the first place they would not employ sufficiently able inspectors, and in the second place they would certainly not do their duty thoroughly, or they would do it doubly in some instances.[41]

Two of the witnesses went even further. Mr A. Potter, a Northumbrian manufacturer, and Mr B.S. Newall, chairman of the Gateshead Town Improvement and Sanitary Committee, both suggested that the alkali inspectors, still under the Local Government Board, should be empowered to control smoke as well as noxious vapours.[42]

The case for an Inspectorate run from Whitehall seemed to be conclusive, until, toward the end of the inquiry, witnesses from Whitehall appeared before the Commission. The commissioners were taken aback. They were faced with strong and unanimous opposition to any enlargement of control from the central government. The Permanent Secretary of the Local Government Board, John Lambert, Dr Ballard, an inspector in the medical department, and John Simon, formerly head of the medical department, left no doubt about their implacable opposition to the control of pollution by central government. We have already had a glimpse of the reason for this opposition; it was an example of the doctrinaire attitude of the Local Government Board toward devolution to local authorities, influenced by Simon, who wanted to eject the Alkali Inspectorate from the Board's responsibilities when local government was being reorganized in 1872. The role of central government, according to this doctrine, was 'to throw as much as they well can upon the local authority'.[43] Ballard, perhaps indoctrinated by Simon, told the Commission that it was 'more in accordance with English practice to leave all matters of administration to the local authorities...'[44] Unless thrown on their own resources, the local authorities would 'never learn'. Simon was equally emphatic. The Public Health Act of 1875 delegated to local authorities the duty to control ordinary sanitary nuisances in their areas; there was no justification for detaching any class of air pollution from the general administration of the sanitary law.[45] After all, Simon argued, the prevention of nuisances from copper works and the like was no more scientific than the prevention of smallpox, and *that* was left to local medical officers, advised, if necessary, by experts from central government.[46]

The only concession that Ballard and Simon were prepared to make was to suggest that if numerous local authorities were unable to employ inspectors of the right calibre, then fewer, higher, local authorities might be appointed for the purpose; either combined authorities (for which the Public Health Act, 1875, made provision) or county authorities of some kind.[47]

This polarization of opinion—on one side the local authorities, the industrialists, and the alkali inspectors in favour of central control, and on the other side the Local Government Board opposed to it—perplexed the commissioners. They mulled over the problem for some time. Experience, they concluded, 'does not encourage the belief that the ordinary local authorities can be depended upon to enforce laws which affect the manufacturing interests of their most important constituents.'[48] As to devolution to combined authorities, that had some virtues: it would, for instance, oblige those who benefited from the abatement of noxious vapours to contribute to the salaries of the inspectors (for the local authorities would have to pay part, at any rate, of the costs and this would give them an interest in making sure the inspections were efficient). Yet there were strong arguments in favour of extending the existing system of central inspection. Such a system would ensure uniformity in standards all over the country; it would enable inspectors to be deployed in any part of the country to enlarge their experience and to prevent those ties of intimacy and suspicions of partiality which might prejudice the judgement of a local inspector; and inspectors in a centralized system would pool their knowledge of the processes under their control and so secure a higher level of good management of pollution.[49]

Although unanimous in their belief that inspection could not be entrusted to 'officers appointed by the ordinary sanitary authorities', the commissioners did not feel they could disregard the Local Government Board's opposition to centralized control. So they sat on the fence over this question, offering the options of centralized control or control by a consortium of local authorities. They did, however, recommend that the head of the Inspectorate should be directly under the Local Government Board, with a laboratory and some assistance, and that local inspectors should reside at the principal centres of manufactures; leaving it to the government to decide who should appoint and pay them.[50]

The Royal Commission's report amounted to an unequivocal endorsement of the Alkali Inspectorate. The precise structure, size, and functions of the Inspectorate were left for decision by the government, but its continued viability was hardly in doubt.

4
THE DYNASTY IS ESTABLISHED

Cautious legislation

Royal commissions are inclined to recommend strategy and to leave it to others to settle tactical matters. The Aberdare Commission recommended a much more broadly based policy for the control of noxious vapours but they were vague about how control should be administered and silent about how control should be paid for. Such fuzziness about practical details was an invitation to politicians to drag their feet, and the responsible minister, Sclater-Booth, accepted the invitation: the upshot of the Commission's labours, he is reported to have said, was 'to leave us very much as we were'.[1]

But the public conscience had been aroused and it was not long before the government came under pressure to do something. In November 1878, three months after the Commission reported, a deputation led by the 15th Earl of Derby (whose father had raised the whole matter for the first time in parliament, 16 years earlier) came to the Local Government Board to get a promise that measures on the lines recommended by the Commission would be turned into law, and that the pattern of administration, left in the air by the commissioners, would be brought firmly to earth. To that end the deputation suggested that the alkali inspectors, increased in number, should be put under a County Board (the deputation was drawn mainly from Lancashire and Cheshire), and that the expense could be defrayed, in part at least, out of the local rates. As *The Times* put it: 'when the people of the suffering districts declare that they are willing to take upon themselves the burden of the cost, they offer a proof of their earnestness...Any measure would probably be accepted with gratitude if it were certain to suppress a nuisance that has become almost a national scandal.'[2] *The Times* added its support for the suppression of the nuisance, and assured its readers that Lord Derby's strong advocacy was rooted in firm utilitarian soil:

> he is, perhaps, the man in all England who is least likely to propose any measure that would embarrass our manufacturers for the sake of preserving or restoring the beauty of the landscape in Lancashire and Cheshire.

But even this assurance was not enough to persuade Sclater-Booth to act. He stalled again. It would not be easy, he said, 'to persuade the manufacturers of chemicals to submit to very severe restrictions and penalties' until local authorities made better use of the powers they already had to control smoke. A couple of months later, on 5 February 1879, a deputation from the Alkali Manufacturers' Association, whose interests would be served if other polluters were subject to restraint, came to Sclater-Booth to encourage him to proceed. He was still playing for time:

it would undoubtedly be my wish, in any extension of the principle of those Acts [the Alkali Acts], to rest such extension upon the widest possible basis, so that it should be, not a particular provision for the benefit of a particular district or a particular class of persons, but that it should be a provision of law applicable throughout the whole kingdom to all alike; and so that all manufacturers...shall be placed, I will not say in similar, but under analogous requirements as to the conduct of their business in such a way as would inflict the least possible annoyance upon the public out of doors.[3]

Under cover of this evasive reply, the Local Government Board was in fact drafting legislation. In January 1879 the Board had sent a circular to local authorities asking for a return of the works in their districts likely to be affected by an extension of the control of noxious vapours,[4] and on 7 April 1879 a Noxious Gases Bill (No.123) was given a first reading in the House of Commons.

The Bill was broadly consistent with the recommendations of the Aberdare Commission. It provided for control over all classes of works liable to any considerable emission of noxious vapours, based on a graduated scale of requirements: fixed limits where they could be fixed, best practicable means if limits could not be fixed, and simple inspection where no means of control were available. As to administration, the Bill stuck to the proven pattern of a centralized inspectorate. And as to the financing of the inspectorate, the Bill proposed to subsidize this from a levy on manufacturers; the origin of what is known as the 'polluter-pays' principle. Local authorities which wanted an additional inspector could have one, provided they would bear part of the cost of his salary. During a technical hitch in the passage of the Bill opposition built up, seemingly not from owners of the new kinds of works which would be brought under control but from the alkali manufacturers. They took exception to two clauses: one which provided for collective liability when nuisance was caused by more than one offender; the other which empowered local authorities in certain circumstances, and not just inspectors, to prosecute offenders. Before the Bill could have its second reading, the parliamentary session ended.

Early in the following session, on 12 February 1880, Lord Derby was back again at the Local Government Board, leading another deputation to press Sclater-Booth to make a thorough job of the legislation; 'to deal effectively and permanently with the evil' (which had grown worse with a happy revival of trade).[5] A fresh Bill, already drafted, was introduced into the House the next day. It was re-named the Alkali Acts Amendment etc. Bill—a significant choice, for it indicated that the government did not want to break away from the style of scientific public service created under Angus Smith. It made a few trivial changes from the first Bill but kept the clause about collective liability.[6] For a second time the Bill fell by the wayside, through no fault of its own: it succumbed to the early dissolution of parliament in March 1880, and it had to start all over again under Gladstone's Liberal administration, sponsored this time by Sclater-Booth's successor at the Local Government Board, J.G. Dodson.

Under the Conservatives the Bill had faltered twice. The Liberals were soon alerted to the duty expected of them. 'Sir', wrote a correspondent to *The Times* (in May 1880, when the new President of the Local Government Board had scarcely found his way to his new quarters):

> What degree of intensity do our successive Governments mean to allow the smoke nuisance to reach?
>
> For the last three years [*sic*] a Noxious Vapours Bill has been burked at the close of successive Sessions. A working man said to me one day 'Well, we dont so much care about their wars, or the Turks and all that, but they do nothing for us. Why if Lord Palmerston had been alive he would long since have rid us of that smoke which poisons us!'[7]

It was the fogs of the winters of 1879 and 1880 that had provoked this angry correspondent, but it is significant for the theme of our story that already in the public mind the two kinds of air pollution—noxious gases and smoke—were being lumped together.

Thereafter the pressure was maintained in parliament itself: a parliamentary question from W. Egerton (a member of the Aberdare Commission) in the Commons on 27 May and another from Lord Midleton (another member of the Commission) in the Lords on 16 July; both parried by the government with vague assurances. When parliament reassembled after the Christmas recess Egerton and Midleton renewed their pressure. Did Dodson intend, or did he not, to bring a new Bill in the 1880-81 session? The number of people affected by noxious gases, said Lord Midleton, 'could not be less than the whole population of Ireland'—a reference that touched a nerve in the parliament of 1881.[8] Dodson in January could give no firm reply; but by 4 February the Marquis of Huntly was able to give the desired undertaking. Three days later he introduced a third Bill, again with a fresh name: the Alkali &c Works Regulation Bill (No. 29). It followed the lines of the Conservative Bill of 1880, i.e. extension to processes other than alkali works, a centralized inspectorate, registration fees to be levied on the works subject to control—but it shrunk from the main decision recommended by the Commission and endorsed by the Conservatives, namely, provision for inspection of a wide range of works emitting noxious vapours, even though means for abating the vapours were not yet available and even though inspection would not have involved any authority to control the emissions. It was self-evident to Angus Smith (whose part in these protracted negotiations we shall discuss later) that the only way to make progress toward comprehensive control was to give the inspector at least the right to observe the uncontrolled vapours at source and to stimulate research into their abatement. The Conservatives had not jibbed at this proposal, but the Liberals, perhaps because they were committed to retrenchment and probably because they were more sensitive to the interests of industry, fought shy of such an embracing measure. The point was made by the Marquis of Huntly:

> where the Government were not prepared to adopt a test regulating the escape of noxious vapours, or any practicable means for restricting them, such inspection would be unnecessarily vexatious to the manufacturers and expensive in practice.[9]

Another omission from the Liberals' Bill was control of coke ovens (a notorious source of acidic gases) on the ground that they emitted also black smoke which could be controlled (though it was not) under the Public Health Act of 1875. The Opposition attacked these pusillanimities: they smacked of compromise, they failed to be comprehensive, they disregarded the labours of the Commission. The Liberals replied extolling the merits of gradualism. Lord Kimberley

> imagined that every government would be glad if it could make all its legislation comprehensive; but it was generally found that in all matters involving considerable interference with trades and business a comprehensive measure was practically impossible. Parliament proceeded, as a rule, and he thought wisely, step by step, building up a system by means of experience...He had been at the seats of great manufactures in this country, and he had certainly felt great pain at the devastation caused by the gases evolved in the manufacturing processes. It was a melancholy thing to see those compelled to live in such an atmosphere; but in this matter of legislation they must be prepared to make gradual progress.[10]

The Times was on the side of a bolder solution. The Bill, wrote a leader-writer (17 February 1881) 'its history, its contents, and its omissions, are but a poor testimony to our progress, public spirit and enlightenment. It...is avowedly tentative; it leaves out at least a dozen nuisances which it is not yet prepared to encounter...'[11] Lobbies joined in protest against the timidity of the Bill. The Manchester and Cheshire Association for the Prevention etc. of Noxious Vapours and the Standing Committee on Health of the National Association for the Promotion of Social Science passed resolutions and proposed amendments. These made little impression upon Mr Dodson, who minuted on one of them: 'Ackge and promise consideration':[12] (the brush-off-courteous in Whitehallese). Attempts in the House of Lords to bring coke ovens under control were defeated. The Bill went to the Commons, where the criticisms of it were repeated, and it was bogged down by Irish members filibustering with taunts that it was a 'Landlords' bill'.[13] Sclater-Booth, now in opposition and therefore able to support a bold and comprehensive approach without having to face the cost and inconvenience of carrying it out, declared that the justification of the Bill as a measure of public policy was destroyed by the narrow limits it had set, and especially the exclusion of a general authority to have under surveillance, even if not under control, the more notorious noxious emissions into the atmosphere.[14]

The Alkali &c Works Regulation Act, 1881, limped into the statute book in August 1881, three years after the Royal Commission had reported.[15] It did little to lift the frustrations which blocked Smith's aspirations but it was in fact a cautious endorsement of his strategy for controlling industrial air pollution.

Before we follow Smith's activities after the Act of 1881, let us turn back to ask what part—if any—he played in the three years of shuttlecock in Whitehall and Westminster, among the drafts of Bills, deputations, and disputes over noxious vapours.

The short answer is that he played very little part. He did, of course, give evidence to the Aberdare Commission and this evidence undoubtedly influenced the Commission's recommendations. But over the drafting of the Conservative Bills of 1879 and 1880 and the Liberal Bill of 1881, he was kept at arm's length. It was not that Angus Smith did not wish to be involved: it was that he was innocent of the rudiments of political guile. During these critical years when governments were deciding what action to take Smith flits in and out of the scene, deferentially making constructive proposals based on his intimate knowledge of the chemical industry, and uncomplainingly resigning himself to nothing more than formal acknowledgements of his painstaking efforts. In November 1880, when the Liberal government was working on a fresh version of a Bill, Smith wrote to the Secretary of the Local Government Board:

> I have ventured to propose a new Act to be substituted for the present Alkali Acts. I do not consider it one for a perfect condition of things but it is in my belief an advance not injurious to any interest...I leave it respectfully for your consideration.[16]

The tone of this letter is not that of a scientific expert already taken into partnership by the administrative head of the Local Government Board! But with the alkali trade Smith had established a partnership: shortly after the advent of the Liberal government the alkali manufacturers took the initiative of inviting him to meet them to discuss new legislation. He had first to seek instructions from the Local Government Board:

> As I have not been informed of any movement on the part of the Government regarding this subject I have written [to the alkali makers] to say so, adding however that I shall be happy to attend a meeting if it is desired subject to the wish of the Board.[17]

The Liberals' Bill was introduced in February 1881 and did not become law till August. During the intervening months it seems likely (very little evidence remains in the records) that Smith kept pestering the Local Government Board's Secretary ('I fear I have written on [the Bill] so often as to tire you...') with suggested amendments to tighten control by the best practicable means, to include chlorine among the gases to be controlled, to schedule *emissions* as well as *processes*.[18] At head office his masters might be forgiven for thinking him to be politically naïve and rather a bore. But no one could question his devotion to the subject and his willingness to help. After one of Smith's notes to the Secretary of the Board, ending with the words: 'I should be glad to know the movements that I may be able to render you aid...',[19] the Secretary did have the good grace to send him two copies of the Bill as it had come from the House of Lords (on its way to the Commons), with a request for any further observations which might occur to him.[20] This was on a Monday; by the following Friday Smith returned the draft Bill with comments based on a meeting he had had with the alkali trade on the previous day.[21] Smith was invited to confer at the Local Government Board,[22] and it is likely that a couple of amendments were made to the Bill in accordance with his advice.

Once again Smith must have been disappointed at the final wording of the Act. It fell far short of the comprehensive surveillance of noxious gases which he had hoped to get. He took the disappointment loyally and without complaint; experience had taught him that making laws was much more difficult than doing science. 'The man who knows fully how to legislate does not show himself to exist, and I believe does not exist...' was the glum comment he made in his annual report for 1879, after the first attempt had been made to draft a revision of the Act.[23]

However, it was the slow maturation of Smith's ideas, as much as the wording of the revised Act, which determined the evolution of policy in the 1880s. These ideas permeate the somewhat woolly correspondence he had with the Local Government Board. Chief among these was the idea that control should be based on gases emitted and not processes that produced the gases: 'Any work whatever', ran his proposed draft, 'giving off any of the following gases into the atmosphere to such an extent as to become offensive, shall be brought under the operation of this Act and the limit of amount of gas shall in all cases not otherwise specified be "the best known method".'[24] This idea was rejected; and as for Smith's formula, whatever he might have intended,[25] there is a subtle difference between a 'best known method' and the 'best practicable means': the one depends only on the state of technology; the other—the word 'practicable'—subsumes other considerations: what is economic, what is consistent with the financial state of the works, the age of the equipment, and so on. There were difficulties also about the use of the phrase 'best practicable means'. If you set a standard (the 0.2 grains per cubic foot for hydrochloric acid, for example) is this deemed for ever to be the best practicable means? Or should the Act contemplate advances in science which would justify setting a stricter standard?

If the inspector finds that one works can control emissions to less than 0.2 grains per cubic foot, should he now demand that other works follow suit? The alkali manufacturers were apprehensive about this and a clause which would have implied it was duly amended.[26]

The Inspectorate responds

Angus Smith was now (1881) 64 years old, but he threw himself with undiminished vigour into the expanded duties covered by the new Act. Despite its excessive caution, the Act did give the inspectors a great deal more responsibility and the Local Government Board matched the responsibility by arranging an increase in staff and salary. Only one new post had been added to the Inspectorate between 1864 and 1881; in 1877 Smith had been given the help of an assistant, W.S. Curphey, who was himself to fill Smith's position from 1910 to 1920. But following the passage of the Act the Treasury sanctioned the appointment of four more inspectors and an additional assistant for Smith. Smith now assumed the title of Chief Inspector and the Inspectorate was restructured to comprise four full inspectorships (which were filled by the former sub-inspectors), and four sub-inspectorships (the newly created posts).[27] Smith's salary had already been reaised to £800 a year in 1878, shortly after his appointment as Joint Inspector under the Rivers Pollution Prevention Act of 1876, and it was now increased to £1000 (still less than that of the Chief Inspector of Factories). The incremental scale for sub-inspectors was fixed at £300 - £400, and that for inspectors at £500 - £550, but Fletcher and another long-serving inspector were awarded special increases. For Fletcher and for Smith the rises were a belated response to repeated representations. Each of them had to press his own case with embarrassing importunity.[28]

The new Act came in to force on 1 January 1882. In the first eight months of operation, the number of registered works rose from about 150 to over 900.[29] Among the works subject to control by the best practicable means were those making chemical manure, gas liquor, nitric acid, sulphate and chloride of ammonia, and chlorine. Emissions from sulphuric acid works were subjected to fixed limits. In addition, two notorious sources of nuisance for which there seemed no adequate technique for control—salt and cement works—were to be inspected with a view to control by Provisional Order when this became practicable.

Smith's staff of inspectors and sub-inspectors now numbered eight, deployed in eight districts. They were immediately overworked, due to the great increase in the numbers of works they had to visit. 'The strain upon us is very great', wrote one of the inspectors from the Midlands.[30] Smith's assistant, Dr Blaikie, fell ill, 'presumably aggravated if not caused by exposure when doing work for the District of the Inspectors who required aid. One of them had said that Inspection was slavery.'[31] Smith himself had to carry a double burden: that of the extended Inspectorate and his new post as an inspector for river pollution as well. Characteristically, Smith saw this as an opportunity rather than a chore. In 1880, before the additional work under the new Act fell on his shoulders, he made a suggestion to the Secretary of the Local Government Board which foreshadowed faintly the firm proposal made three generations later to have a unified pollution inspectorate.[32]

> My wish is to consolidate the Departments of air and water so far as the chemistry is concerned and to report them together, i.e. formally combining the chemical departments of the alkali or vapours act and the Rivers Pollution Prevention Act—a small change.[33]

The suggestion, like so many others which flowed from Smith's incessant pen to the Secretary of the Local Government Board, was ignored.

The two problem-cases that Smith had to tackle were cement and salt. Smith's response to the intractable problem of cement works[34] introduced another seminal idea into the work of the Inspectorate, which has become a very important principle in Britain's present environmental policy. 'It would be very unfair', wrote Smith, 'to make a general law fixing the meaning of a nuisance to be the same in all conditions. Why should a manufacturer established in a desert part of the country be treated like one in a crowded thoroughfare? Or when no one complains, or, rather, when no one is hurt, why should the mere formality of keeping a law be observed?'[35] In modern dress, this flexible attitude to pollution control is the principle that there should not be uniform, blanket, standards for the discharge of wastes into air and water; the standards should be varied according to the capacity of the environment to disperse and dilute the pollutants and—for water—according to the use being made of the river or lake or estuary. It is a principle strongly disputed in some quarters (e.g. it is contrary to the EEC environmental policy). But applied judiciously, it does enable a country to deploy, in the places that need it most, the limited resources available for the protection of the environment. The principle must not, of course, be confused with the famous distinction made in 1879 by Mr Justice Thesiger: 'What would be a nuisance in Belgrave Square would not necessarily be a nuisance in Bermondsey.' The resident in Bermondsey is entitled to as salubrious an atmosphere as is the resident (or, nowadays, the office worker) in Belgrave Square. But the Mersey below Manchester is not 'entitled' to as strict consent conditions for sewage discharge as are, say, stretches of the Wye where trout and salmon flourish; this is the kind of flexibility which Angus Smith advocated when he came to examine cement works. His policy was to compel cement works to abate their objectionable discharges on a sort of sliding scale from relative tolerance to complete prohibition, depending on the amount of nuisance the discharges were causing. This remained the policy of the Inspectorate until as late as 1935, when electrostatic precipitation came into use and a best practicable means could be prescribed.

With salt works now open to him for inspection, Smith was again tempted to exceed his brief. Salt works were as notorious for dense smoke as they were for noxious vapours, and for a long time Smith had been interested in smoke. Here was his chance to do something about it. He was called upon to prescribe for the emission of sulphurous and muriatic acid gases from salt works; perhaps he could prescribe for black coal smoke as well.[36] Processes had recently been developed for the distillation of coal. To encourage these processes Smith mooted the idea of setting up coal distillation plants adjoining salt districts. But it was the familiar experience: no encouragement came from the Local Government Board. By the time Smith came to draft a Provisional Order to cover salt works he had damped down his enthusiasm. In the 'Remarks' attached to his proposals he wrote: 'I should have been glad to advise a clause relating to smoke, but the question is not quite ripe yet although I consider the present mode of burning coal wasteful, and opposed to our abundant knowledge.'[37] Early in 1884, with the Order in draft, he reiterated his disappointment:

> I was hoping that the best practicable means for burning coal at salt works could be introduced. It would be of immense benefit to the country. The number of ways of doing it is very great and it only requires a certain authoritative beginning in salt works and from them it would spread.[38]

Even as we write (1981) the gradual convergence of policies to control noxious vapours and smoke is not yet complete.

When, in 1880, Angus Smith pressed the Local Government Board to make some provision for his retirement, he made it clear that this was merely to contemplate a distant contingency. At the age of 63 he wrote: 'I have no idea of retirement, being in good health.'[39] Three years later he was convincing himself, as workers so dedicated are apt to do, that he was irreplaceable. 'I am hoping not to rest much' (this in his 67th year) 'till after the Session as no one else can take the reins in hand.'[40] In 1884 he fell ill. He went to Colwyn Bay, in Wales, to recuperate, but he could not abdicate. The draft Order for salt and cement works was still not finished and he continued to work at it. The Provisional Order went forward, but Smith did not live to see it presented to parliament. He died at Colwyn Bay on 12 May 1884, as his 20th annual report, already drafted, was going through the press.

Smith's twenty years in the Inspectorate were often solitary, often undervalued, dogged by parsimony of administrators and timidity of governments. He was often disappointed but never soured. The imprint he made on the service has remained in essentials unchanged to this day. Over the succeeding thirty-five years, from 1884 to 1920, the Alkali Inspectorate was directed in turn by three men each of whom had received their training at the hands of Angus Smith. He had founded a dynasty.

5

THE 'POKEABLE, COMPANIONABLE FIRE'

The lobby to dispel domestic smoke

We have seen already how Angus Smith was tempted to ask in 1883 for his Inspectorate to have jurisdiction over smoke,[1] but prudence prevailed and he admitted that 'the question is not quite ripe yet'. We left the politics of smoke enjoying a temporary success under Palmerston in 1853. Let us now return to the subject, to discover how Smith's use of the word 'quite' was a lapse of misjudged optimism.

After Palmerston's initiative the social conscience in favour of smoke abatement was sufficiently aroused to enable smoke clauses to be included in sanitary acts passed in 1858,[2] 1866,[3] and in the famous Public Health Act of 1875,[4] which consolidated the efforts of two generations of reformers to reduce squalor and disease among the poor who were crowding into industrial cities. The smoke clauses in the Sanitary Act of 1866 were mandatory upon local authorities, not merely permissive. Nuisance authorities were obliged to proceed against those in trade and industry who discharged 'Black Smoke in such Quantity as to be a Nuisance' and other kinds of smoke unless it could be shown that the best practicable means had been used to abate it. The Act was largely the inspiration of John Simon, who had been the central medical officer of health since 1855. Simon's great contribution to environmental policy was to create a repugnance to dirt. Breathing smoke was (as one man put it) a 'second nature' to city dwellers; they were resigned to it and, since smoke spelt employment, they were not inclined to bite the hand that fed them. Simon, in retirement at the end of his career, records his indignation at this attitude:

> There are immense masses of our population...who endure without revolt or struggle the extremities of general *Smoke Nuisance*;... accepting, as if in obedience to some natural law, that their common life shall in great part be excluded from the pure light of day...by an ignoble pall of unconsumed soot; and hardly murmuring, in their self-imposed eclipse, that their persons and clothing and domestic furniture are under the incessant grime of a nuisance which is essentially removable. For rich and poor alike...there must be sufficient refinement of taste to abhor even minor degrees of dirt, and to insist throughout on the utmost possible purity of air and water...[5]

Brave words; but the smoke nuisance remained; indeed by the 1870s it had grown worse. There were three reasons for this. The first was that smoke clauses in some local Acts were so weakened by exemptions as to be impotent. The second reason was that even when the police were diligent enough to track down offenders (as they were in the metropolis of London) magistrates were lenient in the fines they imposed. Under Palmerston's Act of 1853 the minimum fine for a first offence was 40 shillings but by the 1880s courts were commonly inflicting penalties ranging from 10 shillings to one penny.[6] The third reason—and it is the theme of this chapter—was that abatement of smoke from steam engines was more than matched by increase in smoke from domestic fires.

No one had ventured to invoke state interference with smoke from the homes of British citizens. In 1880 it was reckoned that there were 597 285 homes in the metropolis, with 3 583 000 fireplaces, and in 1881 seventy-two houses a day were being added.[7] So by the year 1880, when we take up again the campaign to abate smoke, the emphasis had shifted from industrial smoke in the north, conveniently remote from parliament, to domestic smoke in London, persistently visible from the windows of the House of Commons.

Even in these days of mass media it may be a book, like Rachel Carson's *Silent spring*, that ignites the public conscience over some environmental hazard. Just such a book was *London fogs* by the Hon.F.A.R. Russell (Lord John Russell's son), issued toward the end of 1880 and widely quoted for a decade afterwards.[8] All Londoners had experienced fogs but it was Russell's diagnosis that gave meaning to their experience. He noticed something that is still evident in the way people subjectively assess risks. If quite a small number of people are killed in one event in one place (say seven in a railway accident) it makes a much greater impact on the public than a far greater number of deaths dispersed in time and place (such as the 7000 or so persons killed in Britain every year in road accidents). Russell made a similar observation for deaths due to London fogs. A London fog, he wrote, 'performed its work slowly, made no unseemly disturbance, and took care not to demand its hecatombs very suddenly and dramatically'.

In December 1873 there occurred one of the thickest and most persistent fogs of the century. *The Times* for 12 December had a leading article about it, declaring how it paralysed trade, inflicted a loss of time and money, and suffocated cattle on show at Islington. There was no mention in *The Times* leader of hazard to human health; on the contrary, the poor beasts at Islington 'may be considered to have been the chief sufferers'. It was Russell's important contribution to demonstrate that cattle were not the chief sufferers. From the Registrar General's statistics he calculated that during the week of the 1873 fog deaths in London exceeded the average by 700, of which 500 could be attributed to the fog. Late in January 1880 there was another visitation of fog; in the three weeks ending 14 February 1880 there were 2994 additional deaths in London of which some 2000 were attributed to the fog. These figures demonstrated that fog, in its undramatic way, was as ruthless a killer as cholera was. 'Such are the results', wrote

Russell, 'more fatal than the slaughter of many a great battle, of a want of carefulness in preventing smoke in our domestic fires.' That domestic fires were the chief cause, Russell deduced from the fact that some of the worst fogs occurred on Sundays, and Christmas Day 1879 was spent in 'nocturnal darkness' although no factories were operating.

With a thoroughness not matched again until the 1950s, Russell speculated on the cost of damage done by fogs in London. Materials, from fabrics to stonework, suffered. 'The sitting figures...on the north side of Burlington House might, but for their European garb, be taken for Zulus.' The energy London received from the sun, he reckoned, was only about two-thirds of that amount falling on country areas. People had to live in the suburbs to escape fogs and this obliged them to spend money commuting to London every day. Property values in London were depressed by fogs. In a lecture he gave eight years after his book was published, Russell calculated that the annual cost of damage due to fogs was of the order of £5.2 millions, equal to the wages of 100 000 labourers.[9] It is ironic that two great comforts of Victorian life—the coal fire and the water closet—were both indicted for murder: the one through fogs which exacerbated respiratory diseases; the other which carried typhoid and cholera through sewers into drinking water.

Russell's suggestions for abating smoke from domestic fires are in a surprisingly modern key: encourage the use of coke and stone coal (anthracite) by putting a bounty of two shillings a ton on stone coal and a tax on bituminous coals; tax all new fireplaces of uneconomical design; tax cooking ranges not adapted to consume their own smoke; organize hire-purchase schemes so that people can buy well designed stoves; let the Metropolitan Board or some other public authority purchase the gasworks, to supply cheap gas and coke.

Between the 1850s, when smoke in London prompted Palmerston to act, and the 1880s, when Russell's book was published, the number of days of fog per annum in London had almost trebled. Pollution from industry was undoubtedly alleviated to some extent over this period by the laws controlling smoke and by the Alkali Acts. Pollution from the domestic hearth had got steadily worse; the quantity of coal brought to London more than doubled between 1862 and 1882.[10] Political action to abate industrial discharges had proved difficult enough; political action to abate pollution from private homes posed social problems of quite alarming dimensions. In an attempt to tackle these problems the smoke-abatement lobby came into being and played an important part through the rest of the nineteenth century.

There had, of course, been smoke-abatement lobbies before, notably the one led by the militant vicar of Rochdale.[11] But in the 1880s the movement was organized in a much more professional way. It started in October 1880, when a Fog and Smoke Committee, under the auspices of the National Health and the Kyrle Societies, held its first meeting.[12] The Kyrle Society was devoted to the creation of open spaces in London, part of the compassionate movement associated with the names of Miranda and Octavia Hill to bring beauty into the lives of the poor. The National Health Society's president was Ernest Hart, a surgeon and propagandist for sanitary reform. He became chairman of the Fog and Smoke Committee, recruited to its membership some very distinguished people, and guided its activities with statesmanlike skill.

The strategy of the Fog and Smoke Committee was very different from that used by Mackinnon in his solitary struggle against smoke in the 1840s. Before bringing anything to parliament the committee decided to encircle the problem from four directions. It appointed a group to study the state-of-the-art of fuel technology; it asked a lawyer to report on the state of the law on smoke prevention; it enlisted public interest by organizing an exhibition of equipment and design for the abatement of smoke; and it invited some factory owners to confer with them as to how smoke could be abated with the minimum of interference with the manufacturer's interests. Its whole approach was a model in the art of lobbying.

After securing what was, for those days, good press coverage (even its meetings were reported in *The Times*) it began to apply tentative prods of pressure. It sent a circular to the vestries and district boards in the metropolis, reminding them that local authorities already had power to deal with nuisances caused by any 'mill, factory, dye house, brewery, bakehouse or gasworks' which did not, so far as practicable, consume their smoke. Someone saw to it that this circular was noticed in *The Times.*[13] The Committee then persuaded the Lord Mayor of London to allow it to have a meeting in the Mansion House. Again, press coverage was good. *The Times* announced the list of speakers (which included the president of the Royal Society and the First Commissioner of Works) and on the Monday after the meeting (10 January 1881) carried a full account of the occasion.[14] It must have been a heartening event. Notables like the Earl of Aberdeen and the Dean of Westminster were present; there were letters of support from Lords Derby and Aberdare; and Mrs Gladstone, who had evidently been invited, sent belated apologies from Downing Street and added: 'The nuisance of smoke is, indeed, horrible. Mr Gladstone and I often speak of it, and I may express our own feelings most truly in saying how thankful we should be if wise steps be taken so as to bring the matter forward in a practical manner.' If allowances are made for the circumlocutory caution which a prime minister's wife has to display, this was quite an encouraging message. Again, the committee saw to it that this message appeared in *The Times.*[15]

The First Commissioner of Works came to the Mansion House meeting briefed with the government's policy; his speech was a bland mixture of pious hopes and bureaucratic caution. He hoped the time would not be far distant when they would have restored 'the atmosphere of London to its early purity, the blossom to our London roses, and the bloom to the cheeks of our London children'; but he then kept his hopes at a politically safe distance by saying that he would 'deprecate any hasty attempts to legislate', and over domestic smoke he held out no expectation of legal enforcement. It would not be wise, he said, 'to attempt to interfere by any legislation. They must rather trust to persuasion and example and inducements...he did not see the means of persuading the enormous mass of householders to use the smokeless coal unless it could be distinctly proved to them that there would be economy in the change.'[16]

This was a hint to the committee not to rush its fences. It contented itself with a statesmanlike decision that the investigation and testing of various smokeless fuels and appliances 'should precede any application for amendment of the existing Smoke Acts, or for new legislation in regard to smoke from dwelling houses'.

It soon became apparent to the Smoke and Fog Committee that the intractable problems in the cure of domestic smoke were not in technology or economics: they were in sociology. There were already stoves on the market which would burn coke or smokeless anthracite coal. These fuels cost at that time no more than ordinary coal. But the householder, eager enough by now to suppress smoke from factory chimneys, was unwilling to forego his own cheerful fireplace for the silent and colourless emanations from a closed stove; nor, by all accounts, were servants anxious to exchange the daily chore of brushing up cinders and laying new fires for the cleaner job of tending an anthracite stove. (At one meeting of the Fog and Smoke Committee Mr Owen Thomas offered 'to send a couple of Welsh maids to town to show London maids how to burn smokeless coal in stoves suitable for domestic purposes'.)[17] So the top priority in the Committee's campaign was to overcome the inertia of householders.

This was the purpose of the exhibition which opened in South Kensington on 30 November 1881. It was launched with an impressive ceremony in the Albert Hall.[18] Public interest was such that the assembled company, which included Princess Louise and the Marquis of Lorne, was 'numerous enough to fill the entire area of the hall'. The exhibits included a great variety of improved grates, stoves, and furnaces for domestic and industrial use, and an assortment of fuels. More than 100 000 people visited the exhibition, including the Prince of Wales and the Empress Eugénie. Special arrangements were made to bring parties from distant parts of London and from the provinces. When the show closed in February 1882 it was reopened in Manchester. There was good coverage in the press. The most fortunate and timely support came from the environment itself: on 18 January and again on 3–4 February there were awful fogs. London traffic came to a standstill; cases in the police courts had to be deferred because witnesses lost their way; theatres were practically deserted and those plucky enough to venture out to see a play found fog obscuring their view even inside the theatre.[19]

The exhibition was not merely a showpiece; the exhibits were subjected to quite sophisticated tests. The heating power per pound of coal consumed and smoke produced was worked out. Measurements were taken of the amount of unburnt carbon, the hydrocarbons, and the weight of solid matter intercepted on filters. On these and other criteria awards were made for the most efficient stoves in each class.[20]

All this publicity implanted the idea that smoke was an avoidable nuisance which ingenious designers were working vigorously to dispel. There was tangible evidence that some people were responding to the exhibition's message; one firm which displayed a cheap stove at South Kensington sold 14 000 of them in the succeeding two years.[21] But expectations that there was an acceptable cure for domestic smoke were disappointed. There was no doubt that closed stoves burning anthracite could dispel the nuisance, but there was also no doubt that Englishmen did not like closed stoves. The smoke-abatement lobby had to consolidate itself in order to have the stamina for a long campaign. This it did in 1882, by translating the informal Fog and Smoke Committee into an incorporated National Smoke Abatement Institution (NSAI). The lobby had by now mobilized very impressive support: the Duke of Westminster as president; other members of the nobility; distinguished professional men including the president of the Royal College of Surgeons; and men who had already made history in the applications of science to social reform: Edwin Chadwick and Lyon Playfair.

At this stage the lobby might have been tempted to press for some action in parliament. It wisely resisted any such temptation, for the case for bringing domestic smoke under the law was still weak. The man in the street was still not willing to barter his open fire for cleaner air and fewer fogs. Even if he had been willing, the problems of converting stoves and grates, and supplying millions of tons of smokeless fuel to the metropolis, were far from solved. Any premature precipitation of the issue in parliament would surely have led to failure. Instead, the lobby played for time by passing a resolution 'that the Council of the National Smoke Abatement Institution be requested to urge upon the Government the desirability of appointing a Royal Commission...' to pursue the matter.[22] The hope was that this could be floated on the wave of rising expectations such as *The Times* discerned and described in an important article on 6 November 1883:

> The pall of smoke, which every year becomes denser and more choking, cannot be thinned without interference with millions of offending chimneys. Happily, the force to remove obstructions is materially not unequal to them. Of late years it [London] has been adding to its material resources of aggregate wealth and combination the

moral incentive of a vigorous collective determination of Londoners to render their metropolis less depressing and less squalid. The standard has risen, and is rising daily, by which they reckon their claims upon the community they constitute to cheer as well as support existence. Their expectations were very moderate formerly. They were satisfied if the neighbourhood of masses of their fellow human beings was not actively noxious. They are beginning to require now that it should be... positively agreeable.

The saga of Lord Stratheden and Campbell

Six months passed, and the government made no response to the request for a royal commission. It was at this stage that an enthusiast for clean air made what, judged by hindsight, was a tactical error. On 26 May 1884 Lord Stratheden and Campbell presented in the House of Lords a Smoke Nuisance Abatement (Metropolis) Bill.[23] The gist of the Bill was to empower any local authority in the metropolis to make byelaws 'for prohibiting or regulating the emission of smoke from any building within their district'. The Sanitary Act of 1866 already contained a sub-section prohibiting the emission of black smoke but this Act had an exemption: 'not being the Chimney of a private Dwelling House'. It was this exemption which was dropped in the Bill presented by Lord Stratheden and Campbell. He told the House that it was 'invidious to proceed against furnaces, which at least contribute to the public wealth, and leave untouched the ordinary fireside, which only burns for the advantage of the circle who surround it'.[24]

It was pretty clear that Lord Stratheden and Campbell had not been briefed by the National Smoke Abatement Institution to introduce this Bill; the Institution's policy was to consolidate its case and to buy time during which it could strengthen public opinion. It was an embarrassment to the government, obliged to concur in the need to abate smoke but aware of the formidable practical difficulties in the way. And the House of Lords could hardly go on record as opposing a motion to rid London of fogs. It was no surprise that the Bill passed unopposed at the first reading. At the second reading the grounds for opposition began to emerge.[25] Was this not an invasion of the Englishman's home? Emotive phrases like 'the defence of liberty and property' crept into the debate. The government spokesman, Lord Dalhousie, by a technique only too familiar in politics, proceeded to obstruct the progress of the Bill by a diversion: it would be premature, he said, because the government had it in mind to pass a Bill to reorganize the pattern of local government in the metropolis. In any case, it was too late to pass the Bill into law during that session of parliament. Nevertheless, Lord Stratheden and Campbell scored a qualified success; although the Bill did not survive the second reading it perished with honour: 31 lords voted in its favour and only 17 against.

Lord Stratheden and Campbell knew that a private Bill stood little chance of success. His intention was to keep the issue alive in debate. So, undaunted, he introduced a similar Bill in the next session, on 19 March 1885.[26] For a second time the House of Lords allowed the Bill to proceed unobstructed to its second reading, on the assurance that there was no intention to carry it further. He came back with a third Bill in March 1886.[27] This sank without trace.

Meanwhile the anti-smoke lobby, baulked of its move to get a royal commission, had no option but to welcome the efforts of Lord Stratheden and Campbell, in public at any rate. In 1886 the National Smoke Abatement Institution decided it had better resume the initiative. Fortified by a particularly nasty fog in December 1886 which lifted the mortality rate to the level caused by a bout of cholera, the Institution sought an interview with the Home Secretary.[28] This time it was civil servants who put up the obstacles. 'There would seem to be little use in a Deputation coming', was the advice they gave the Home Secretary,[29] followed a few days later by a more urbane version: 'Shall they [officers of the NSAI] be told that though the subject is one to which S of S [Secretary of State] attaches importance there is so small a prospect of legislation this session that S of S does not consider that any useful purpose would be served were he to give himself at the present time the pleasure of receiving a deputation.'[30] This sugar-coated rebuff was endorsed by the Home Secretary and the NSAI informed accordingly.[31]

On 14 March 1887 the pertinacious Lord Stratheden and Campbell launched his Bill for a fourth time.[32] On this occasion the smoke-abatement lobby stood by him. The Duke of Westminster, president of the NSAI, took a leading part in the debate. The government, seeing that pressure was now building up dangerously and sensing that it might soon be pressed to take some action, tried, through the Home Office spokesman, Earl Brownlow, to tranquillize the House by saying that the matter 'would not be lost sight of'; but this was not enough. At the end of the debate the House did what it would have been better to have done at the beginning: it agreed to refer the matter to a select committee.[33]

The Select Committee met only twice; it interviewed only four witnesses; and it produced a singularly unimpressive report.[34] After this half-hearted inquiry the Bill came back to the House of Lords slightly amended to strengthen its provisions and too late in the session to receive thorough consideration. It was debated on 1 August[35] and immediately confronted by an amendment by the Earl of Wemyss:

> That before the law for the prohibition of smoke is extended to private dwellings, it is desirable that the purpose and intention of the existing Acts be more carefully carried into effect either by their amendment or by their better administration.

This was the signal for the opponents of the Bill to bring out all their weaponry. Of course they were in sympathy with its objects. Of course they were concerned about the welfare of the poor who could not escape from the fogbound city. *But*...and there followed a formidable list of impediments. It would be a 'hateful system of espionage' to invade the privacy of homes in order to inspect stoves. Public opinion was not ripe for such a measure. London had 'got on' despite fogs; indeed a sanitary expert had asserted that the carbon made London air less deleterious than it might otherwise have been. The Bill, said the Lord Chancellor, was 'ill devised, clumsy, and inconvenient'. The motion to take the Bill to the next stage was lost by 30 votes to 12 and the procrastinating amendment from the Earl of Wemyss was duly agreed to.

Endlessly patient, calmly reasonable, consistently courteous, Lord Stratheden and Campbell presented his Bill for the fifth, sixth, seventh, eighth, ninth, and tenth time between 1888 and 1892.[36-41] There were minor variations, but he stuck to his purpose with the tenacity of a spider spinning a web which is constantly being destroyed. His seventh, eighth, and ninth Bills did contain some novelties, but these only attracted fresh criticism. Local authorities were to be obliged to proscribe 'opaque' smoke, and the penalties for infringement were to have regard to the defendant's 'means and condition in life'. In the debate these provisions gave Lord Salisbury an opportunity to exercise a tasteless sense of humour. Interpretation of the word 'opaque', he said, 'would give infinite pleasure, amusement, and occupation to Her Majesty's Courts of Justice'. As for the Courts having power to take into account the 'means and condition' of defendants: 'Does that mean', he asked, 'that a Duke may make more smoke than a Baron?'[42]

The last episode in this saga of perseverance in the public interest came in 1892 when Lord Stratheden and Campbell, on 11 February, presented his Bill for the tenth time. In his new version 'opaque' smoke had become 'black'; a change which would not have satisfied John Bright.[43] Lord Salisbury did not challenge this change but he had no intention of letting the Bill go through. He was pressed by Viscount Midleton to set up a royal commission; this he refused to do, after indulging in some elephantine humour at the expense of royal commissions in general. But Viscount Midleton's questions did draw from the Prime Minister the real reasons for his resistance even to inquiries into smoke abatement. 'The difficulty', he said, 'is political...I do not know whether my noble Friend thinks he would ever get parliament to pass such a measure [it was to make the use of smokeless fuel obligatory in London] or whether he would get the English people to obey it if it were passed'. If this were to be made obligatory for Londoners, it would 'condemn the population to go for ever...without ever seeing a fire with a flame in it; and I do not think that, for the sake of avoiding an occasional inconvenience, grave as it is, for a certain number of days in the winter, people would condemn themselves to a flameless fire all the winter through'.[44]

Lord Stratheden and Campbell made his last appeal, at the second reading of his Bill on 3 March 1892. He was, as always, courteous and conciliatory.

The Duke of Westminster supported him, with ominous figures for the increase in frequency of fogs (there were 93 in 1887, and 156 in 1890). Reluctantly the Prime Minister allowed the Bill to go to a Standing Committee; the Committee considered the Bill on 5 April and adjourned until 10 May. The Bill never reappeared.

Lord Stratheden and Campbell's stoical campaign to rid London of smoke was over. Not quite over: it had a touching postscript. He died on 21 January 1893. On 6 June 1893 a question was asked in the House of Commons about the progress—or lack of it—in the Smoke Nuisance Abatement (Metropolis) Bill. Mr H.H. Fowler replied that the will of the late Lord Stratheden and Campbell carried the following provision:

The control of my Bill on smoke abatement, which has been sanctioned by the House of Lords and gone since through accurate revision, I venture to leave to the Duke of Westminster and his coadjutors.[45]

Surely the strangest legacy one man ever left another! Its redemption date was 63 years later, when the Clean Air Act, 1956, finally included domestic fires in its jurisdiction.

How was it that a politician endowed with such indefatigable patience, supported by a lobby so well organized, so reasonable in its demands, so distinguished in its roll of supporters, and with its message reinforced over a hundred times a year by fogs, failed to get a Bill to abate domestic smoke on to the statute book?

The superficial answer is that the timing was wrong. Lord Stratheden and Campbell failed to distinguish between the enlightened aspirations of a few hundred enthusiasts at a public meeting in the Mansion House and the indifference of a million householders in the suburbs. *The Times*, which had on the whole supported efforts to abate smoke, got it right when it wrote, a couple of days after the tenth Bill was introduced in the House of Lords, that a fog was better than a cheerless hearth and the intrusion of inspectors (not in fact required by the Bill) into the homes of Englishmen.[46]

But this is only a superficial reason for the failure. We have to inquire why the timing was wrong. If the State is to intervene to curb pollution, three conditions have to be satisfied. There has to be a practicable technology for abating the pollution; the law must be capable of surveillance and enforcement; and—since surveillance and enforcement are never by themselves sufficient—the law must be acceptable to the public who are called upon to surrender a little freedom in exchange for a common benefit. For certain noxious vapours and for smoke from steam engines these conditions could have been met, though we have seen how cautious governments were to extend the Alkali Act to new processes, and how careful they were to provide escape routes for offenders against the laws to abate smoke. For domestic smoke the circumstances were more complicated. No one denied that homes could be heated by smokeless fuel, burned in closed stoves or even in suitably designed open grates, provided there was gas to ignite the fire and some form of efficient draught. But such sources of heat did not look or feel like open fires. Even the scientific journal *Nature*, quoting the opinion of a distinguished sanitary engineer, Sir Frederick Bramwell, agreed: 'We are strongly of the same opinion...that we must have an "open, pokeable, companionable fire".'[47]

The obstacle to smoke abatement was not primarily technological: it was sociological. Sociological obstacles to progress are every bit as intractable as are technological obstacles, and we still know far too little about how to tackle them. It would be correct to say that in the 1880s there was already a best *known* means to abate smoke from domestic hearths. But it was not a best *practicable* means, for the problem of how to persuade people to do without visible flames had not been solved.

It was over the surveillance of a law covering domestic smoke that the most emotive arguments were used. It is not necessary to enter a house in order to see smoke coming out of the chimney. But much was made about 'an army of police watchers' and the intrusion of inspectors into private homes. A more serious question was how the law was to be administered in the Courts. The Bills put before parliament gave no guidance about the interpretation of their technical provisions. For example, all fires smoke when they are first lit: for how many minutes was such smoke to be tolerated? And how were the escape clauses—the defence that 'reasonable measures' had been taken to prevent smoke—to be operated? Who was to be held responsible for infractions: the ground landlord (who might have leased the house for fifty years), or the tenant (who might be coming to the end of an annual tenancy)?

The reformer's path was littered with obstructions such as these, some of them substantial, others mere quibbles. Beneath the prevarication there was, of course, a much deeper motive for resistance against laws to abate domestic smoke. Under common law most domestic smoke could not be proved to be harming anyone or anything; to control it would be to make something illegal which had been legal for centuries. This was a forfeit of liberty not to be made lightly.

There is an object lesson in the episode we have described in this chapter. It is the need to understand not just the technology of an environmental problem, but the mechanics of slow maturation of public opinion about it. *The Times*, looking back on the decade of the 1880s, put it well:

> We must...endeavour to stir up public opinion to the point of action by implanting a conviction in the public mind that civilisation itself is retarded by the toleration of nuisances that can be removed and of dirt that never ought to have been created... we are not without hopes that the day is not far distant when public opinion will insist that London must somehow be relieved of its canopy of gloom and its hideous vesture of dirt.[48]

6

GASES OR PROCESSES?

A new Chief Alkali Inspector

While parliament during the 1880s was fending off pressures to control smoke in London, the industrial areas in the north of England were contending not only with smoke, but with noxious vapours. The Alkali &c Works Regulation Act of 1881 did cautiously extend the authority of the inspectors over a few more processes. Angus Smith, the founder of the Inspectorate, died still in harness, soon after drafting memoranda to the Local Government Board in an attempt to bring escaping gases from salt and cement works under control.

On the very day *The Times* carried the announcement of Smith's death (13 May 1884)[1] the Secretary of the Local Government Board minuted in his department: 'Prepare appt of Mr Fletcher as Chief Inspector'.[2]

Alfred Evans Fletcher was recruited in his thirties by Angus Smith at the very beginning of the Alkali Inspectorate. He was Smith's devoted, but not uncritical, disciple. For twenty years, from 1864 to 1884, he had been based on Liverpool, so he had unrivalled experience of the growth of chemical industry in Britain. He came from a well-known Nonconformist family; his father was founder and first head of Denmark Hill School, at which the sons of many distinguished Nonconformist families were educated. Fletcher was sent to Berlin to complete his schooling; he then worked from 1845 to 1847 on the railways; and completed his education, specializing in chemistry and mathematics, at University College, London. He joined Angus Smith after a disappointing attempt to develop the aniline dye industry in England.[3] (Aniline dye was one of the first examples of a failure to follow up a triumph of pure research in Britain with its industrial exploitation; it was in Germany that the British invention was successfully exploited.)

We have already met Fletcher in this story, at one time pressing for a rise in his pay, at another giving evidence to the Aberdare Commission. His policy was to continue the policy of his predecessor, but he brought to the work a more forceful personality, a more lucid mind, and a greater determination to enlarge the Inspectorate's responsibilities. Over eleven years, until his (reluctant) retirement in 1895, he kept the confidence of manufacturers and won the confidence of those we would now label environmentalists. The only group he could not persuade to his views were the civil servants, but even with them he had a greater influence than Angus Smith was able to exert in his time.

We have already explained how the revised Act of 1881 was a disappointment to Angus Smith. Soon after Fletcher had settled into the job, he began to press for a revision. His annual report for 1887 contained a critical review of the Act and an announcement of his policy for its reform.[4] Its title was an anachronism, its scope inadequate for the needs of contemporary industry. The emission of hydrochloric acid from alkali works was under control; other noxious gases were beyond control not through lack of technology but through deficiencies in the law. Thus in the recovery of sulphur the foul and poisonous gas sulphuretted hydrogen was released, but it was immune from the requirement that the best practicable means should be used to suppress the gas. The law was drawn so narrowly that manufacturers using, as distinct from making, some of the proscribed gases, were beyond control because the processes of manufacture had not been scheduled under the Act. Fletcher put forward in his annual report two clear proposals: the first, that a revised Act should list the gases to be controlled, from whatever source they came, and not (as hitherto) the processes liable to emit them; and the second, that the prescription of fixed emission limits (e.g. the 0.2 grains per cubic foot of hydrochloric acid in alkali manufacture) should be replaced in all cases by the more flexible formula of best practicable means. He reinforced this latter point by saying that, owing to improved technique, the emissions of hydrochloric acid were, in fact, already much below 0.2 grains. He repeated (with a slight variation which perhaps is a misprint) the important principle he had given eleven years earlier in evidence to the Aberdare Commission:[5]

It would, I think, be found that the injunction to use the 'best practicable means' for preventing the emission of noxious gases, would prove an elastic bond [*sic*] ever tightening as chemical science advanced and placed greater facilities in the hands of the manufacturer.

His conclusion ran as follows:

If this view of the subject is adopted a comprehensive measure for controlling the escape of noxious gases might be so framed as to depend mainly on two clauses, the first defining a noxious gas as one which gives cause for complaint, or referring to a schedule in which such gases are named; the second, declaring that the best practicable means must be adopted to prevent the emission of such a gas.

Stonewalling in Whitehall

Fletcher had given notice, to the public as well as to the Local Government Board, that his aim was to persuade parliament to replace the Alkali Act with a general noxious vapours Act. The next four years of his time in office were used, over and above his daily duties, to campaign for this end. He had the smoke-abatement lobbies on his side. Industry, with some exceptions, did not seem hostile to the idea. The intransigents were in Whitehall. In

January 1889 Fletcher discharged a memorandum to the Secretary of the Board, reminding him of the points made in the annual report for 1887.[6] The reaction in the office was predictable: had not similar clauses in the 1879 Bill been dropped in the Bill which became the Act in 1881? However, the Secretary advised the President that Fletcher should be asked to prepare an informal draft of the provisions he would like to have put in a Bill, but he added this advice to his political master:

> My own view is that the Board could not propose legislation to apply to every gas 'which gives cause for complaint'. The definition should be much more precise.
>
> Neither should I think it desirable, as regards certain specified gases, to do more than bring within the provisions of the present Act those gases for which provision is required but which are not now within the Act.[7]

The President, busy at this time reorganizing local government and not wanting any more on his plate, agreed with this advice. Fletcher prepared a detailed list of the gases to be indicted as 'noxious'. This formidable list was enough to confirm the President's resistance to any further action: 'I cannot', he minuted, 'contemplate further legislation in the present session.'[8] It took more than this kind of rebuff to put Fletcher off. He replied eleven days later, telling the President through the Secretary that the proposal to control gases rather than processes was 'but reverting to the instructions given to the Royal Commission on Noxious Gases of 1876. They were instructed to enquire into the "working of Factories from which sulphurous acid, sulphuretted hydrogen, ammoniacal and other vapours were given off".'[9] His reply was brought to the President's attention. The papers were then 'reserved' for six months, till November 1889.

Meanwhile, Fletcher was deploying his allies. Promptly, in November 1889, came a memorial from the Guardians of the Barton Regis Union in Gloucestershire, seeking an extension of the Alkali Act to include lead smelting works, spelter works, glass works, and potteries, with specific reference to the need 'admitted' in Fletcher's annual report for 1887, followed by a strong memorandum to the Board from Fletcher himself.[10] This memorandum produced ominous figures for the release of sulphuretted hydrogen in works which recovered sulphur from alkali waste. In one of these alkali works alone, it was expected that the weekly production of sulphur would be 180 tons, 'involving the production and manipulation of the enormous quantity of 4½ million cubic feet sulphuretted hydrogen gas'. Under the terms of the Act, this process was outside the control of the alkali inspectors. Under cover of this special case Fletcher repeated his plea for control by specifying gases instead of processes—something which would involve no change in the spirit of the Act. This was followed by a meeting with the Secretary who tried to persuade Fletcher that it was undesirable to unsettle the Act so soon after 1881. On the contrary, wrote Fletcher when he returned home to deal by letter with his interview, the short time elapsing since 1881 was rather in favour of amendment; moreover there was growing pressure for amendment (three more memorials had been received, including one specifically praying for the scheduling of gases); and further, there was an urgent need for inspection of works producing metallic fumes of arsenic, lead, and zinc.[11] This was forwarded to the President of the Board with the comment: 'Mr. Fletcher you will observe is strongly in favour of a Bill of much wider operation'; to which the President replied: 'I cannot undertake legislation of so wide a character in the meantime'.[12]

The failure to control sulphuretted hydrogen, however, was becoming a scandal, and Fletcher did get the President's initials of approval on a request to bring under the Act 'Those processes of manufacture in which sulphuretted hydrogen is produced'.[13] The request, even with this approval, stuck at the level where it had to be prepared as a Bill. The official concerned queried such a comprehensive wording. It would be open to question, he wrote, 'whether on account of a comparatively small emission...very important classes of works (e.g. cement works, gas works) shd be made subject to registration and inspection'.[14] Fletcher was therefore asked to make a list of the main processes which would come under the proposed amended clause. He listed four, and this simply confirmed the Secretary of the Board in his belief that it was safer to specify processes of manufacture and not gases. Fletcher was told that any amendment to include sulphuretted hydrogen would be in the form of additions to the list of registered processes. He protested, but, obliged to submit for the time being, he specified not four, but seven processes which—if processes *had* to be specified—should be on the list; adding again his protest at this decision:

I cannot refrain from again reminding the Board that such a schedule is essentially cumbrous, incomplete, and indefinite. It will need supplementing and correcting from time to time, while on the other hand clauses placing all manufacturing processes which involve the generation of sulphuretted hydrogen, hydrochloric acid etc. under supervision would be inclusive and definite. I have so fully stated my grounds for this opinion that it is unnecessary now to repeat them. See my Report under Alkali Act for 1887, pages 15-21, also subsequent minutes.[15]

The Board proceeded to draft a clause to include the processes listed by Fletcher. Drafts were referred to him and were returned not only with the comments he was asked to make but also with a dogged reiteration of the case for 'gases' rather than 'processes'. 'I would again express a hope...' in one minute; 'I trust the Board will pardon my insistence...' in another minute. He even taunted the Board about the fear that, if gases were the basis of control, and not processes, some works might be included in error. 'Can any such work be named', he challenged, 'or any manufacturing operation from which are evolved any of the noxious gases mentioned in the schedule proposed in my letter of November 25 last which should not come under a noxious vapours Act?'[16] The President was unmoved: 'I adhere to the view already expressed that particular works should be specified'.[17] The Bill was duly drafted and Fletcher, for the time being, subsided. On 20 June 1890, a Bill which would have widened the Inspectorate's powers to cover more processes was ready; only to receive from the President, on 21 June, the dampening response: 'We can't introduce any more Bills of this kind this Session'.

This delay in a Bill thoroughly unacceptable to Fletcher gave him another opportunity to bring pressure to bear on the Board. He now laid down a veritable barrage of complaints. On 28 November he forwarded a complaint about fumes from zinc works in the Western district. Three days later he reported that the alkali makers of Liverpool thought the amendments in the deferred Bill 'very desirable'. Five days after that he drew the Board's attention to a complaint from the Clerk of Widnes about 'frightful escapes of sulphuretted hydrogen' due to the adoption of the Chance sulphur recovery process. On 31 December, doubtless stimulated by Fletcher, the Manchester and Salford Noxious Vapours Abatement Association renewed their earlier memorial asking for legislation on the lines of Fletcher's 1887 annual report. Five days after that an association of sanitary authorities in the Liverpool region sent the President of the Board a memorial asking for protection against nuisance from alkali works in Widnes, with further reference to Fletcher's report of 1887.

The barrage successfully softened up the Board. On 27 January 1891 Fletcher was invited to discuss the question of legislation. The Bill was taken off the shelf and refurbished in the light of a further memorandum from Fletcher. On 23 March it was circulated to the Cabinet. Again there was disappointment (for even this feeble revision would have been better than nothing): the Bill was put aside to make way for the LCC Finance Bill and the Postponement of Elections Bill.[18]

Fletcher had loyally suppressed his conviction that the revised Bill should be designed to control specified noxious gases and not specified industrial processes, and he had co-operated fully in the drafting of clauses with which he basically disagreed. Nevertheless, now that the Bill was to be delayed again ('one of the innocents—yet unborn—that must be sacrificed for this Session' was the President's final ruling on the draft),[19] Fletcher used the platform of his annual report for 1890 to revive his idea of a general noxious vapours Act. His case seemed to him so reasonable that he was clearly puzzled by the obtuseness of politicians:

It has been pointed out in former reports that it would be very desirable if the Act were so framed that new processes of chemical manufacture could be brought within its scope as soon as it was ascertained that certain noxious gases were liable to be discharged from them...It would not then be necessary to describe each separate process of manufacture and indicate the method by which this gas is generated. The fact that such a gas was present and liable to escape into the air would be sufficient to bring the operation within the scope of the Act.[20]

This report, published in 1891, was followed by a renewal of pressure from outside, doubtless stimulated by Fletcher. The memorialists from Liverpool followed up their previous complaint about the nuisance from Widnes by a plea for drastic action. The stink of sulphuretted hydrogen was 'exceedingly offensive'; inhabitants on the outskirts of Liverpool found the 'stench at night...occasionally so strong as to awaken them from sleep'. Schedule the noxious gases themselves, was their demand to the Local Government Board.[21] The memorial was sent to Fletcher for comment. It was an invitation to him to return to his theme:

I have often urged that the Alkali Act should be extended in the direction indicated in this memorial and trust that it may now be possible to take this step.

The short amendment of the Act as now printed wd serve as a palliative...but the more general recasting of the Act on the lines indicated wd be greatly to be preferred.[22]

It was in vain. The Secretary of the Board minuted to his President 'I presume that you will adhere to the view that the Bill should be proceeded with in the form in which it has been drafted' to which the President replied 'I think so';[23] and after further pressure, slightly more testily: 'I see no reason to alter the view previously expressed.'[24] With the persistence of an Aberdeen terrier, Fletcher kept up the pressure to the end. He badgered the Secretary of the Board; the Secretary replied in a weary and irritable tone; Fletcher responded with unrepentant explanations:

to describe separately each such process [he was pressing for the extension of the Bill to some more metallurgical operations] would be difficult or well nigh endless, but they might be included in the phrase 'processes for the treatment of metallic ores whereby sulphurous acid is evolved'. This however would probably be thought too general and to invade the principle [i.e. of processes, not gases to be specified] it is desired to maintain.[25]

At long last the Bill came to the House, on 7 April 1892. It attracted comparatively little comment and passed into law on 27 June.[26] The Act is a two-page document which did nothing except to add 13 works to the schedule, including those that produced sulphuretted hydrogen; and to exempt from the Alkali Act certain works making salt from rock-salt. The Liverpool lobby made a last minute attempt to have the Bill withdrawn in favour of a Bill with the wider powers Fletcher wanted.[27] But the politicians knew (and admitted) that a Bill with such wider powers would never get through. Progress by 'Minute Particulars': this has been a characteristic of the evolution of policy for the control of pollution in Britain. Fletcher considered this style of progress to be sluggish and inefficient, but in our view it is one reason why this policy—compared with the policies of some other countries—has proved workable.

Fletcher, speaking through his annual reports, made no pretence of disguising his dissatisfaction with the amended law. It was 'much to be regretted'; every year would reveal some additional process which might with advantage be brought under the Act—and now new regulations would be required every time such a process was disclosed.[28]

Fletcher's achievement

Our description, so far, of Fletcher's career as Chief Inspector must have created the impression that he spent most of this time nagging at the Local Government Board. This impression would be incorrect and we must now restore the balance. Despite the narrowness of the Act, Fletcher greatly broadened the scope of the Inspectorate. His significant contributions to the history of a clean air policy in Britain were first, that he consolidated the concept of best practicable means as a tool to control pollution, rather than as an escape route for offenders; and second, that he boldly went outside his sphere of authority to tackle the intractable and neglected problem of how to abate smoke. Let us now look at his achievements in these two activities.

Angus Smith regarded best practicable means, (or, as he sometimes called it, 'best known means')[29] as a temporary expedient to empower inspectors to require noxious gases to be abated even though the degree of abatement could not be quantified. Fletcher, as we have seen, believed that the very flexibility of this phrase was an advantage, for it would automatically ensure that legislation kept pace with advances in technology. Under the extended legislation of 1881 there were, by the time Fletcher succeeded Smith, no fewer than 914 scheduled works subject to the 'best practicable means' requirement, and only 127 alkali works subject to a fixed emission limit. So how did Fletcher decide just what were best practicable means?

He adopted two principles: one, to fix the best practicable means only after full consultation with the manufacturers; and two, to enforce compliance with the agreed guidelines by persuasion and not, except as a last resort, by prosecution. Here is an example of his technique.

One of his first tasks was to devise a best practicable means for abating the escape of chlorine from chambers containing bleaching powder. He drew up a draft code, six rules, which he submitted for discussion to meetings of manufacturers in Liverpool and Newcastle. The meetings were (to quote his annual report) 'numerously attended' by manufacturers, who were sympathetic to Fletcher's code and who 'aided much' in reaching a consensus about it. There was agreement that 'some means must be provided, and be kept in constant use, for absorbing the chlorine which remains in bleaching powder chambers after the powder is finished. The air finally discharged into the atmosphere or into a flue or chimney must not contain more chlorine that 2½ grains per cubic foot.'[30] Early in 1886

Fletcher was able to report that in view of the gratifying results he felt he could now urge all makers of bleaching powder to adopt one or other of the methods 'which have already been shown to be practically available'.[31] But he still held his hand about setting a presumptive standard for emissions. Later that year he tightened the 'elastic band': a circular was issued to the makers of bleaching powder telling them that inspectors had been instructed to make tests frequently on the understanding that the emission of air containing more than 5 grains of chlorine per cubic foot would be considered a contravention of the Act. The response was such that Fletcher was able to offer in his next annual report a graceful tribute to the trade:

It is pleasant to review a journey which has led to a successful issue. It gives me also an opportunity of showing how willingly the manufacturers have joined in the effort to remedy an evil when its extent was clearly shown to them, submitting in generous self-restraint to a little harassing which seemed necessary in order to cause motion, shall I say, to excite chemical action.[32]

The best practicable means, evolved tactfully in this way for new processes, became under Fletcher's guidance 'a standard which can never grow antiquated, nor can it be oppressive; neither on the other hand is it so loose as to be ineffective'.[33]

Part of his success with manufacturers was his realism. Back in 1877 he had told the Aberdare Commission:

In treating with a manufacturer, it should never be forgotten that his works are established not to condense gases, not to suppress vapours, but to make money, and all fancy processes which interfere with this result stop the works, and thus defeat their own object.[34]

He was criticized for the scarcity of prosecutions under the Act, but he defended himself firmly.[35] He was not administering a police Act. His inspectors were not collecting materials on which to base prosecutions; their job was to assist industry, not to impede it. Their instructions were not to be overbearing or offensive. The result was that the alkali inspectors earned respect and even encouragement from the managers whom they were inspecting. So highly did one manufacturer 'value the visits and examinations of the district inspector that he would consider it worth his while to contribute 100*l* [pounds] annually, rather than the office should be abolished'.[36]

Respected by the Local Government Board (despite his persistent advocacy of a change in the principle of the Act), welcomed by manufacturers (despite the constraints he was obliged to put upon them), Fletcher went on to make some progress in getting support from a third constituency: the general public. Here he went clean outside his terms of reference, and tackled the troublesome problem of smoke. It was not in Fletcher's nature to make excuses for exceeding his powers, but he was able to rationalize his attitude by saying that smoke from coal contained sulphur dioxide, an acidic noxious vapour as harmful to vegetation and property as hydrochloric acid.[37] He went into action against what he called the 'giant' nuisance in his second report, published in 1886. His justification was the enormous quantity of coal being burned in salt works, which had been brought under the Alkali Act by a Provisional Order in 1884, though only for muriatic acid.[38] A new general heading appeared in his annual reports: 'Black Coal Smoke'. He admitted he had no statutory authority to confront this nuisance, but he was convinced that the nuisance could be controlled by proper stoking and proper design of boilers and that the time had come for him 'to write on the subject and to remind the public that when they desire it, the nuisance can be made to cease'.[39]

Fletcher was right to lay the responsibility on the public. It was already an offence under the Public Health Act of 1875 for industries to emit black smoke; the law was not properly enforced because it was too difficult to win prosecutions. But the laxity of magistrates reflected the apathy of the public. So Fletcher devoted a great deal of his time trying to ignite public opinion. It is a tribute to his skill in public relations that the annual reports of the Chief Alkali Inspector—hardly to be classed as entertaining literature—were more widely read by the lay public and quoted in the general press in Fletcher's time than they were in the time of Angus Smith—and indeed, more than they have been ever since.

So, while the Lord Stratheden and Campbell was tending his infertile Bills for clean air in London, Fletcher prepared the soil in Manchester to support lobbies for smoke abatement in the North. In 1882 the exhibition of equipment to abate smoke, put on in South Kensington in the autumn of 1881,[40] was re-opened in Manchester. But its influence had evaporated, and it was left to Fletcher, in his annual report for 1886, to ram home the contrast between the air of London, where Lord Palmerston's Act had suppressed a good deal of smoke from factory chimneys, and the towns of Lancashire and Yorkshire, where smoke from the chimneys of cotton mills and some iron works was doing far more damage to the quality of the environment than were noxious gases from works registered under the Alkali Act.[41] Fletcher's propaganda, supported by grim data for the increased mortality in the smoke-laden towns of the North, prompted a revival of interest in smoke abatement. The Manchester and Salford Noxious Vapours Abatement Association decided to encourage the technology of smoke abatement by arranging for the testing of appliances for preventing smoke. They established a committee comprised of impressive eminences: peers, bishops, mayors, and a galaxy of scientists. Fletcher was appointed chairman of the executive committee. A distinguished engineer was selected to carry out tests. Trials proceeded over five years, from 1891 to 1896, and the results of a battery of tests, ranging from level of smoke and efficiency of combustion, to cost in wear and tear, were published in 1896.[42]

The conclusions in the report are remarkable for their triteness. They contain scarcely anything which had not been familiar to Mackinnon's Select Committee, some fifty years earlier; they made no advance on the assessment given by de la Beche and Playfair in 1846. Still the only technological improvement of any significance was the more robust mechanical stoking device available in the 1890s. Cities with smoke clauses in their Towns Improvement Acts had power to enforce smoke abatement but there was still not enough stiffening of public opinion to compel magistrates to enforce the law.

Fletcher's final excursion outside his statutory boundaries was to launch a propaganda sortie against domestic smoke. He lived in London and must have been familiar with the ten Bills floating in and out of parliament during his tenure as Chief Inspector. He began by referring, under his annual heading 'Black Coal Smoke', to a primitive—but apparently effective—central heating system he had installed in his own home. It was 'a stove placed in the basement of the house, so arranged that an ample supply of gently warmed air shall pass from it and permeate all the passages and apartments'.[43] The cost of warming his whole house this way was only £3 for the whole winter, whereas his neighbours' coal bills over the same period were not less the £25.[44]

On the principle that only by endless repetition can an idea be driven into the heads of the public (and of politicians), Fletcher filled his 'Black Coal Smoke' column year after year with pleas, warnings, and protestations. One of the baneful effects of black smoke, he wrote, was to separate the poor from the rich, 'the toilers who are tied to their daily toil from their wealthier employers and others whose means and occupation allow them to leave the smoky city at sundown and spend their evenings and holidays in the clearer air of the country'—a separation, he added, 'at the root of much social evil'.[45] Some people—perhaps the Local Government Board itself—had objected to his inclusion of the smoke question in his annual reports. But, he protested, the inspectors were up against the smoke question all the time; many of the complaints they received against registered works were due to black smoke, not to noxious gases.[46]

It is a matter for conjecture whether or not Fletcher's campaign against smoke had any success. He himself thought it did. It may well have been his influence which weakened the efficacy of best practicable means as a defence against polluters when they were brought to court for failing to control smoke from steam boilers. 'It is now becoming more fully acknowledged on the magistrate's bench', he wrote, 'that the evil is to a large extent preventible; manufacturers have therefore been pressed to improve their systems of coal burning, so as to inflict less evil on their neighbours...'[47]

Like Angus Smith, Fletcher was reluctant to retire. It was a tribute to the esteem in which he was held by his employers, despite the way he harried them over the deficiencies of the Act under which he worked, that on three separate occasions his appeal for an extension of his contract beyond the customary retiring age of 65 was strongly backed by the Local Government Board and conceded (albeit reluctantly) by the Treasury.[48] H.H. Fowler, when President of the Board, stiffened one draft to the Treasury thus (the italics denote his red ink insertions):

> Mr. Fowler...is *strongly* of opinion that it is of *the greatest* importance in the public interest that Mr. Fletcher should continue in his office until the provisions of the new Act are brought into full operation.[49]

This was a remarkable testimony to a scientific inspectorate created tentatively only thirty years before; the work primarily of two men: Angus Smith, the pioneer, and Alfred Fletcher, who was Smith's colleague from the beginning.

During those thirty years the Inspectorate's work had grown more complex and more exacting. New chemical industries were springing up every year and each one presented fresh problems in the disposal of waste. Fletcher had moved to London shortly after his appointment (Angus Smith had kept Manchester as his headquarters throughout his career as Inspector). He found some years later that the polluted air of London was bad for his wife's health and he applied to the Local Government Board for permission to live out of London; permission which was granted only reluctantly by that chronically insensitive and unhelpful body.[50] Fletcher explained that 'my duties keep me much at home. They consist in maintaining the correspondence which arises from reports and letters constantly received from the Inspectors under the Alkali Act and other people; in writing reports, and in directing the work of my chemical laboratory'.[51] So, being obliged by the bureaucratic fall-out which accompanied the growth of the Inspectorate to be tied more and more to his desk, Fletcher sensibly kept his office in his home outside London, and of course was frequently away visiting his inspectors.

As for the inspectors, they still had to work under what were often trying conditions in what would now be called 'unsocial hours'. Thus, in order to bring chlorine works under control the inspectors had to be present to take tests 'from 2 am onwards', for that was the usual time when the chlorine chambers were opened.[52] And the work was still hazardous at times: apparatus for making tests had to be carried up lofty and awkward ladders, and the inspectors were constantly exposed to the noxious vapours it was their duty to control. Nevertheless the inspectors were treated as professionals and they dressed as gentlemen, attired in frock coat and top hat; an impressive sight, as one of them is remembered, descending at a gas works from a horse cab hired at the railway station.[53]

Fletcher retired at the end of May 1895, after eleven years as Chief

Inspector. His links with the Inspectorate remained, for his son Eustace, whom he had appointed as his assistant in 1899, spent his career in the Inspectorate, though he did not rise to the rank his father held. Alfred enjoyed a long retirement, and continued to take 'the keenest interest in the work of his old department'.[54] Notwithstanding the onerous exertions he had sustained in the early days of the Inspectorate, he survived to the hoary age of 93.[55]

7

PROGRESS AND STAGNATION

Consolidation of the Alkali Acts

Fletcher's tenure of the office of Chief Alkali Inspector was distinguished by sustained pressure upon his masters, the Local Government Board, to expand the scope of his duties and to back his Inspectorate with tougher legislation. Fletcher's successor, R.F. Carpenter, gave his masters a quieter time. He was conscientious, reasonable, 'a most pleasant colleague',[1] but—compared with his two predecessors—undistinguished. He had worked in the Inspectorate for thirteen years before succeeding Fletcher. He brought to the office a pious devotion to the doctrine propounded by Angus Smith and an acquiescent conformity to the narrow ambit of the Act under which he operated. Time and again in his annual reports Carpenter paid respectful homage to what had by now become the orthodoxy of the Inspectorate. Thus, he dropped the comments on smoke from coal which had become an annual feature of Fletcher's reports, for smoke was not strictly within his province. He lost no opportunity to declare how co-operative and law-abiding the industrialists were. His chief concern was to conserve the amicable symbiosis which Angus Smith had created between inspectors and manufacturers: the purpose of inspection of premises, he said, was educational, not inquisitorial or punitive. Throughout his fifteen years in office he brought only twenty cases to court for causing pollution. Prosecution he regarded as a 'painful' duty which struck at the roots of mutual confidence. 'It cannot be too strongly pointed out', he wrote, 'that unless this feeling of mutual confidence exists and is cultivated, inspection becomes much less efficient in protecting the interests of the public.'[2] Fletcher would, of course, have agreed with this, but Fletcher's velvet glove concealed a much tougher hand than Carpenter's.

Notwithstanding these fulsome tributes to industry, there was plenty of room for improvement in the control of noxious vapours. Some pollution persisted because there were no techniques for abating it; some persisted because parliament had not given the inspectors enough authority to curb it. Carpenter was willing to go along with legislation as it stood; it is ironic that events drew him into a major overhaul of the laws, establishing reforms which remained the basic statutory instrument for the Inspectorate until 1975, eighty years after Carpenter took office. Let us now describe how this happened.

We know that the Local Government Board had no thought of embarking on fresh legislation so soon after the amendment of the Alkali Act in 1892. Carpenter took up duty on 1 June 1895. In the following June a request came from the Liverpool Association of Sanitary Authorities for action over the nuisance from copper smelting works. The Local Government Board, to whom the request was made, returned the unequivocal reply that it had not in contemplation 'at the present time any proposal for the further amendment of the above [Acts]'.[3] But two years later Carpenter had to ask the Board for clarification over a technical point.[4] This set off a chain of events which ended in a major revision of the alkali laws in 1906. Unfortunately the records for this period are incomplete and we can do little more than speculate about the influences and pressures which linked Carpenter's inquiry with the 1906 Act. The fact that the process took eight years may well be due to Carpenter's reluctance to force the pace of reform. He was content to toe the departmental line and to acquiesce in changes which he did little to influence.

At any rate we know what the technical point was which set the events in motion. In the manufacture of sulphuric acid, acid gases of sulphur and nitrogen are evolved. The 1881 Act prescribes that 'the total acidity of such gases in each cubic foot of air, smoke, or gases escaping into the chimney or into the atmosphere does not exceed what is equivalent to four grains of sulphuric anhydride'.

But: cubic foot of what? The gases as they escape from the chambers where the acid is made? Or the gases after they have mixed with various other emissions and have come out of the chimney? Other clauses in the Act specify this limit as applying to 'any discharge of gas *by* a chimney or flue...' (italics ours). It had been the custom for inspectors to take their samples, not from the top of the chimney, but from the exit of the process chambers. Obviously, if the top of the chimney had to be the sampling point, the permissible concentration, taking account of the dilution by smoke and other things, ought to be different. Some manufacturers were beginning to exploit this ambiguity.

There were discussions at the Local Government Board to decide how to redefine the clauses in the Act to avoid what Carpenter, in a minute on the matter, described as the risk of a lower standard of efficiency (it would be a 'national calamity', he said, if this were to creep in).[5] It was a real risk, for the United Alkali Company was, by 1898, taking advantage of the opportunity to be lax about the emissions of acid gases, under cover of the ambiguity in the law; and this, Carpenter pointed out, was contrary to 'Angus Smith's views as to what was directed by the Act of 1881'. The Board's legal officers confirmed that the existing phraseology was weak, and in April 1901, after the inevitable delay in the preparation and introduction of an amendment to the Act, a single clause Bill was introduced into the House of Lords, concerned solely with the precise point at which tests for acid gases were to be carried out in sulphuric acid works. Carpenter had asked for a couple of other changes to enlarge the cover of the Act, but these were not included.

The Bill passed uneventfully through the House of Lords but ran into difficulties when it reached the Commons.[6] Industry, apparently, had not been consulted. Manufacturers brought pressure to bear on the Local Government Board to amend the Bill, substituting the best practicable means in place of fixed emission limits as the criterion for compliance for certain new processes. This, of course, was a stimulus for other criticisms of the Bill to be made. If the government had come to some arrangement with alkali manufacturers to soften the standard, then was it not time to think about raising standards and to look at the whole Act again? Should its amendments not be 'of a comprehensive character'?[7]

The Local Government Board, thwarted in its attempt to slip a little technical amendment through parliament, found to its embarrassment that it had loosened the tongues of MPs to debate the whole issue of parliamentary control over industrial pollution of the air. Sir John Brunner, a spokesman for the alkali manufacturers, assured the House that

> For many years the alkali trade has been managed by men of very high scientific attainments, who have always treated the Government Department as a friend and not as an enemy, and the manufacturers have been met in exactly the same spirit by the Department for forty years...[8]

To which John Burns (who ultimately piloted the revised Act of 1906 through parliament) replied:

> Judging by the appearance of some parts of the country around the factories carried on by these gentlemen of high scientific attainments, they do not seem to have shown that kindly consideration for their rural and agricultural neighbours which...men of high scientific attainments ought to be capable of displaying.[9]

These tripwires did not precipitate an exhaustive debate on the matter. But they were sufficient to provoke some further amendments (to make control of hydrochloric and sulphuric acid processes more strict), and that was sufficient to delay the Bill until August 1901, when it suffered the familiar fate of most clean air Bills: it was dropped for want of parliamentary time.

This was not the end of the matter and the Local Government Board knew it. It prepared for further agitation from victims of industrial pollution. Farmers in the industrial north-west of England were complaining bitterly about the lack of any control over metallurgical works. The Aberdare Commission, back in 1878, had recommended that these works should be brought under surveillance. Accordingly a detailed Bill, amending and consolidating the Acts of 1881 and 1892, was prepared and (to avoid the complaint made about the previous Bill) this one was deliberately introduced just two days before the end of the session (on 12 August 1903) so that it could be aired during the summer recess.[10] Conferences were held with people who would be affected by the Bill, and the industrial lobby evidently got its own way, for when the Bill was reissued and read a first time in May 1904 metallurgical works no longer appeared in the listed schedule at the end.[11] The Bill was obviously not going to have an undisputed passage through parliament; so it was withdrawn at the end of the session in 1904 and a fresh Bill, introduced in May 1905, suffered the same fate.[12]

By 1906, however, there was an enthusiastic Liberal government in office and when a fresh version of the Bill was introduced, early enough in the session (19 March) to be fully debated,[13] it was successfully carried through by the new President of the Local Government Board, John Burns (whose caustic remarks about 'men of high scientific attainments' had helped to trip up the original one-clause Bill in 1901). It had a surprisingly uneventful passage; surprising because it contained several novelties and because, although it began with no mention of metallurgical processes, an amendment to provide for inspection of these processes was carried in the House of Commons and survived scrutiny in the House of Lords. On 4 August 1906 the Bill passed on to the statute book as the Alkali &c Works Regulation Act, 1906.[14] There it remained, unchanged apart from additions made to the schedules of processes, until 1975.

The 1906 Act was notable for three reasons: (1) It confirmed the government's policy for abating noxious gases, namely to extend control piecemeal, cautiously adding processes to the schedules when there was some prospect that there might eventually be a best practicable means of abating pollution from them;* (2) it consolidated the policy—which Fletcher tried so hard to upset—of designating processes and not the waste gases they produced as the basis for inclusion under the Act. However, the Act did clarify the expression 'noxious or offensive gas' by giving a list of 13 such gases; though the clarification remained equivocal, for a noxious and offensive gas issuing from a process not on the register remained out of reach of the law; (3) it adhered to the three categories of control which had evolved under Angus Smith's leadership, namely fixed limits for a few kinds of emission, the use of best practicable means for emissions where fixed limits would be impracticable, and a third category of emissions subject only to inspection and study, in the hope that techniques might be found to abate them. One important clause in the Act was a description of the meaning—it was hardly a definition—of best practicable means, which marked another step in the evolution of this indispensable formula. It was originally assumed to refer to the design of the furnace or equipment; it was extended to cover

* In the first year of operation of the new Act the number of registered processes increased by 351 to 1821 and the number of registered works by 153 to 1231. (*AR for 1907* (1908)). These included tin-flux works (60), smelting works (44) and others that do not appear in earlier statutes. It greatly increased the responsibilities of the Inspectorate. There was no corresponding increase in the Inspectorate's staff.

the 'efficient maintenance' of the equipment; and in the 1906 Act it applied 'also to the manner in which such appliances are used and to the proper supervision, by the owner, of any operation in which such gases are evolved'.[15]

In addition to all these points, the new Act did deal with the inquiry which had at the outset stirred up the renewed interest in the Act, namely to define precisely where acid gases had to be sampled in sulphuric acid works.

Carpenter, as Chief Inspector, must have taken part in the many drafts which culminated in the Act of 1906, but there is little evidence of his influence and no evidence whatever that he was setting the pace. In fairness to him it has to be said that he was a sick man. A couple of years after his appointment he was suffering from nervous exhaustion; two of his annual reports were delayed by illness; and he had in the end to seek early retirement.[16] His somewhat anodyne tenure ended in 1910; but he had faithfully cherished the covenant created by Angus Smith and he had the satisfaction of seeing that the new Act destroyed nothing of the revered tradition of the service, and indeed enlarged its scope and set a seal of approval on its style. At the turn of the century it could be said with confidence that there had been, since the 1860s, modest and steady progress in the control of noxious vapours from British industry.

Sir William Richmond writes to *The Times*

Over the same period, smoke in the industrial cities had got worse. Even London, which enjoyed a temporary benefit from Palmerston's Act in 1853, had more fogs in the decade 1890–1900 than ever before or since (Fig. 1). Carpenter had turned his back upon this nuisance. How was it that for one kind of air pollution there had been progress and for another kind stagnation?

There is a simple reason. From the very beginning the control of noxious vapours was a sophisticated matter. They were, most of them, invisible. Their presence could not be demonstrated, still less measured, without the techniques of science; and their abatement could not be achieved without considerable skill in chemistry and engineering. This was the main reason for putting the control of noxious vapours under a scientific staff employed by central government, and the reason was amply justified. Smoke and the foul smells from organic decay needed no science to demonstrate their presence, and it was assumed—though quite wrongly—that they could be abated without sophisticated science. So these nuisances were left to be dealt with by local government. People were told (for example, by Mr H.H. Vivian, MP, a South Wales industrialist) 'You cannot have manufactures carried on without suffering these disabilities [he was referring to industrial smoke]: half or two thirds of your incomes is derived...from manufacturing

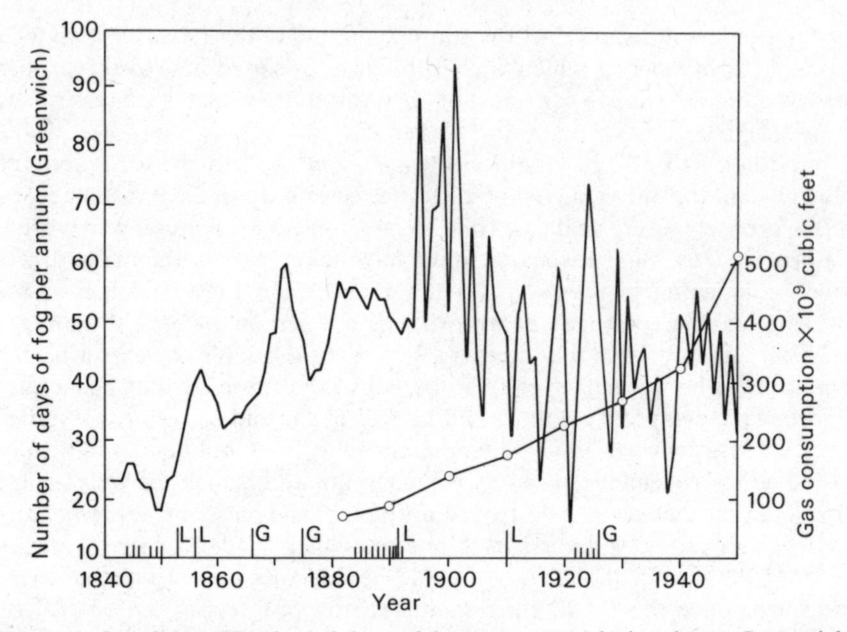

Fig. 1. *Left ordinate.* Number of days of fog per annum during day at Greenwich, 1841–1950. The values from 1924 to 1950 accord with a definition of fog as visibility 'less than 1 km'. The values from 1841 to 1921 do not accord with any specific definition. The data from 1841 to 1950 can, however, be regarded as sufficiently coherent to illustrate the trend. Data by courtesy of the Meteorological Office. Sources: 1841–1894, *Q. J. Met. Soc.* **23**, 287 (1897); 1895–1906. Greenwich Observatory; 1907–1950, Meteorological Office monthly reports.
Right ordinate. Sales of gas by statutory undertakings in Great Britain, 1882–1950, in cu ft $\times 10^9$. Data by courtesy of the British Gas Corporation, Economic Planning Division.
Vertical lines along abscissa. Short lines: Smoke Abatement Bills which were not successful. Long lines: Smoke Abatement Acts passed. G = England and Wales; L = covering only London.

industry, and you must take the rough with the smooth.'[17] There was a benumbed acceptance of black smoke as the inevitable consequence of civilization. There were indeed technological problems to be solved before smoke and dust and grit could be economically controlled; but the chief problem was public apathy. And the only weapon against apathy was the voluntary association of protesters: the anti-smoke lobby. Let us see how this lobby struggled at the turn of the century to quicken the public conscience.

We now pick up the story of smoke abatement where we left it at the end of Chapter 5. Lord Stratheden and Campbell had touchingly bequeathed his smoke-abatement Bill to the Duke of Westminster, and *The Times*, on

23 May 1890, had called for public opinion to be stirred to the point of action.[18]

We resume the story fittingly on 17 November 1898: on that morning a black fog covered London. By ten o'clock it was 'too dark to permit the reading of letters; and, half an hour later, the darkness was comparable to that of a total eclipse of the sun'.[19] Sir William Richmond, the eminent artist best known for his mosaic decorations in St Paul's Cathedral, found himself unable to draw or paint. Instead, he wrote a letter to *The Times* describing the fog as an 'object lesson to inert and indolent ratepayers of London'.[20] The vestries of the London parishes had power under the Public Health (London) Act of 1891 to prosecute offenders but they failed to use their power. Richmond himself, 'by dint of persistent bullying' had forced the Hammersmith Vestry to prosecute in two cases, 'both of which we won'. People, he wrote, should 'club together, and badger their vestries as I have badgered mine'. To this end he suggested that 'a society be formed which shall make it its business to report to the various vestries of London where-ever the infringement of the Smoke Act is observed'. Smoke-vigilantes to patrol the streets! 'The law is clear', he wrote, 'the enforcement of it is all that is needed.'

The matter was not so simple as that. There were serious deficiencies in the law. This did not exonerate the dilatory vestries but it did make successful enforcement difficult. The basis for litigation against smoke lay in section 91 of the Public Health Act of 1875. (This did not apply to the metropolis of London but its essential provisions were embodied in the Public Health (London) Act of 1891). The gist of section 91 of the 1875 Act was to declare as a nuisance:

> Any fireplace or furnace which does not as far as practicable consume the smoke arising from the combustible used therein...and Any chimney (not being the chimney of a private dwelling-house) sending forth black smoke in such quantity as to be a nuisance...

Then followed an escape clause:

> That where a person is summoned before any court in respect of a nuisance arising from a fireplace or furnace...the court shall hold that no nuisance is created within the meaning of this Act, and dismiss the complaint, if it is satisfied that such fireplace or furnace is constructed in such manner as to consume as far as practicable... all smoke arising therefrom, and that such fireplace or furnace has been carefully attended to by the person having the charge thereof.[21]

No wonder local officials hesitated before bringing a culprit to court under these provisions. 'Black' had never been defined in the courts, and 'as far as practicable'—by now a useful tool in the expert hands of the alkali inspectors—was a safe defence in the hands of a clever lawyer against allegations of nuisance from smoke. Not only did these technical difficulties deter local officials; there was also a psychological difficulty, namely that dwelling houses were exempt, and by the 1890s it was common knowledge that the prime sources of smoke in many cities were the kitchen range and the open hearth in millions of homes. Manufacturers, therefore, could claim to be victimized if they were penalized and the worst offenders remained immune. To compound the difficulties, there was a further reason for laxity in enforcement of the law. From 1853 to 1891 enforcement in the metropolis was in the hands of the police, but the Act of 1891 transferred this duty to the sanitary authorities—the vestries and district boards. These bodies were unprepared for this duty and at first had no means of discharging it.

It is a familiar quirk in history that some trivial event seems to trigger off a long and significant episode of social change. Sir William Richmond's letter to *The Times* was just such an event; from it there followed a fresh wave of activism, one which was sustained, even through the disruption of two world wars, until public opinion was ready to welcome the Clean Air Act of 1956. *The Times* reinforced Richmond's letter with a sympathetic leading article on the following day.[22] Day after day, there followed volleys of letters under the heading 'The Smoke Nuisance'. There was caustic criticism of the local vestries and of the newly created London County Council (the LCC) which had power to intervene if a vestry failed to remedy a nuisance.[23] A letter signed 'An aggrieved ratepayer' incited citizens to have vestry clerks prosecuted for neglect of duty. On 24 November, a week after the critical fog, a new ally appeared: a correspondent from the gas industry supported the call for action. For the first time, the smoke abatement lobby was supported by industrial vested interest. This letter to *The Times* put paid to any notion that there was no adequate substitute for the coal fire and the kitchen range. The number of gas fires and stoves on hire from one London company, the letter said, 'placed side by side would reach from Charing Cross to the West pier at Brighton and 50 miles back again'.[24] Over 150 000 artisans' houses had been supplied with gas heating over the preceding five years, and the amenity was brought within reach of the less well-to-do by the installing of penny-in-the-slot meters.[25]

A few days later the response to Richmond's letter began to take shape. Mr Verney, a member of the LCC, called a meeting of sympathizers at his London home. The Earl of Meath (who was an alderman of the LCC) took the chair. The following resolution was moved by Sir William Richmond himself:

> That a society be formed and a committee appointed to consider and report how best the vestries and other municipal bodies can be aided by public support to enforce the clauses in the Public Health Act which relate to smoke consumption.[26]

This was passed unanimously and a committee which included such eminences as the Duke of Westminster, Viscount Midleton, and Rollo Russell—all survivors from the lobby of the 1880s—together with distinguished scientists, Lord Kelvin, Professor Thorpe, and Professor Meldola, was elected to set the society in motion. Richmond became its founding president and Dr H.A. Des Voeux, a practising London surgeon who was to inspire and drive the society for the ensuing forty years, became its honorary treasurer.

The Coal Smoke Abatement Society

So began the Coal Smoke Abatement Society (CSAS). It made a prompt and aggressive start. Little more than a year after the November fog which provoked Richmond's letter to *The Times* the Society reported its first action to awaken vestries to their duty. 'With this object in view', wrote Richmond, 'the society appointed an inspector, Mr. Petty, whose name has since figured largely in the police reports of the public Press—indeed he has become a terror to offenders against the law and to perfunctory vestry-men.'[27] Mr Petty was indeed a terror: in the first six months of his service, from May to November 1899, some 1000 observations were made which led to the reporting of some 500 nuisances, i.e. black smoke emissions of at least ten minutes' duration in one hour. The sanitary authorities took action against 60 of these; 23 nuisances were immediately abated; 16 summons were issued and fines amounting to £200 were imposed. Not surprisingly, Mr Petty's exertions did not endear him to the vestries or to the LCC, and his task was made harder by the derisory penalties exacted in the courts. From 1899 to 1904 the average fine was less than £5. Some offenders were content to go on paying every time they were caught, rather than spend money to abate the smoke. Slowly the local authorities were won over to support the Society and to co-operate with the inspector; with one notorious exception: the borough of West Ham. This recalcitrant body had, by 1907, received 1788 complaints of nuisance. They stubbornly refused to do anything about any of them and, to make matters much worse, the Local Government Board, which should have intervened, took no steps to exercise its powers of coercion.[28]

Through Mr Petty's zeal culprits were brought to court. Through the sympathetic support of the press good publicity was given to the Society. But this success failed to bring about the prime aim of the Society: to enlist practical support from the public. Subscription for membership of the Society was only five shillings a year. It was hoped to build up a modest membership of 4000, which would have brought in an income of £1000 a year. This hope was not fulfilled. Such was the apathy of the public that even by the year 1905 the annual income was only £300. The Society had to appeal for funds to relieve itself of the 'constant and harassing incubus of debt' from which it was suffering.[29]

The incubus of debt was no deterrent to enthusiasts like Richmond and

Des Voeux. To proselytize was, in the long run, more important than to prosecute and, for this, official encouragement was more important than cash. The encouragement came in 1905 when the Commissioners of Works invited the Society to co-operate in a series of tests on open grates entered for competition at the new government buildings in Westminster. Tests on thirty-six designs were carried out by a trio comprised of a representative of the Commissioners, an architect, and Des Voeux from the Society. The results were published in *The Lancet*, a journal which gave sustained support to the Society's efforts.[30]

In London domestic hearths were undoubtedly the commonest source of smoke but they were not the most egregious. Electricity generating stations, industries such as Doulton's ceramic works (uncomfortably close to the Houses of Parliament), steamboats on the Thames, and railway locomotives were the more visible polluters. As long ago as 1846 it was accepted that even bituminous coal could be burnt in steam furnaces without offence, provided the furnaces were properly stoked. The CSAS revived this point by setting up classes for stokers; in 1907 about a hundred of them were being instructed in efficient combustion at the Borough Polytechnic.[31]

Education was the first priority of the CSAS; pressure on vestries to enforce the Public Health Act was a second priority; and the third was 'Where the present law is inefficient to bring about an amendment'.[32] But, like its predecessor the National Smoke Abatement Institution, this new lobby deliberately refrained from pressing for legislation. Any move to amend the law would raise the emotive issue of the control of domestic smoke, and, in dealing with industrial smoke, there was a new incentive which might become more effective than statutes, namely the plain fact that efficient combustion saved money. The idea that it is actually *profitable* to abate smoke was gaining ground, just as it had already been dramatically proved in the alkali industry that profits could be made from the recycling of wastes from soda works. Another incentive more effective than statutes was the convenience and increasing availability of gas. Gas was still, in 1905, more costly than coal but the promoters of gas were overcoming the prejudices against it and boosting its conveniences, some of which were positively hedonistic; as, for example, a 'by-pass controlled by a lever at the bedside, which permits the fire to be called into activity as a preliminary to rising...a luxury which only those who have enjoyed it are fully able to appreciate'.[33]

Frustration in the London County Council

Thanks to the enterprise of the gas industry and the efforts of the CSAS, London air became cleaner in the first decade of the twentieth century. Fogs became less frequent—this is evident in the graph on p. 82—and the amount of winter sunshine rose by about 40 per cent.[34] But London was still a disgracefully dirty city compared with other cities on the Continent. As a preliminary to any move to stiffen the law, the London County Council, under pressure from the CSAS, asked in 1906 for a government inquiry into the whole problem of smoke abatement. This request was received by the Local Government Board with its customary apathy.[35] Then in 1907 there occurred an incident which strongly reinforced the case for amending the law.

The Chelsea Borough Council had sued the Underground Electric Railways of London Ltd. to abate the emission of black smoke from the power station in Lots Road, Chelsea. Mr Curtis Bennett gave judgement in favour of the Electric Railways and attached to his decision some highly prejudiced opinions. London, he said, owed a great deal of gratitude to a company which had caused underground traffic to be worked by electrical power. In his opinion the works of the generating station were as perfect as science could possibly make them. Furthermore he found as a fact that what unscientific witnesses, called by the Borough Council, had described as black smoke was 'in truth and reality only dark brown smoke'. And in token of his prejudices he felt that the case was one in which substantial costs should be allowed. He dismissed the summons with 300 guineas costs to be paid by the Council.[36]

The Chief Officer of the LCC's Public Control Committee had already recommended that the LCC should promote a Bill to strengthen the Act of 1891.[37] Its clauses were crippled by qualifying words: the term 'black' and the phrase 'as far as practicable'.[38] Other weaknesses were that the LCC had no powers under the Act to finance research or education in smoke abatement, nor had they sufficient powers to intervene in the affairs of metropolitan boroughs. Proposals to overcome these deficiencies were drawn up and submitted to the Corporation of the City of London and to the twenty-eight metropolitan boroughs. The response was dismal: only seven authorities wholly approved, eleven approved partly or conditionally, six disapproved, four had no opinion, and one did not bother to reply. Nevertheless the LCC pressed on and by June 1909 it was ready to apply to parliament to delete the word 'black' from the penal clause about smoke and to tighten up the law in sundry other ways.[39]

The proposals, embodied in the LCC (General Powers) Bill provoked immediate hostility. In the House of Commons there was a move to reject the section on smoke *in toto*, and other moves to strike out the clause about the deletion of the word 'black'. The word 'black', said the London Chamber of Commerce, was 'their only safeguard against more or less vexatious proceedings'.[40] In the Local Legislation Committee of the House of Commons supporters of the Bill were confronted by thirteen petitioners objecting to various parts of it. They included the London Underground Railways, the Central London Railway, the London Chamber of Commerce, the Electric Supply Companies of London, the Gas Light and Coke Company, and the South Metropolitan Gas Company.[41] Hearings went on for ten days. The notorious Borough of West Ham objected to any jurisdiction from the LCC outside the county boundary. The smoke section of the Bill emerged in a deplorably emaciated condition. The main amendment which the LCC regarded as of cardinal importance was disallowed; the word 'black' remained on the statute book.[42] All that survived of the LCC's Bill were three amendments: (1) the LCC could take action at the request of a sanitary authority in cases of special difficulty; (2) the LCC could take action against smoke pollution by a sanitary authority itself; and (3) the LCC could spend up to £500 a year for investigations of smoke nuisance. This feeble flicker of legislation received the royal assent in August 1910.

The Coal Smoke Abatement Society was created to enforce smoke-abatement laws made for the metropolis of London, but in December 1905 it widened its horizon by holding a national conference and exhibition, organized jointly with the Royal Sanitary Institute.[43] Over a hundred delegates from municipalities and engineering societies attended. A timely overture to the event was a severe fog. Delegates from the north, where the atmosphere was commonly murkier than in London, used the occasion to propose that there should be consolidation among the many smoke-abatement societies. A notable contribution to the conference was a paper by Professer Cohen of Leeds. He presented a dismal picture: 20 tons of soot produced daily in the city, ten hundredweight of it falling within the city boundary; up to a half of the sun's light cut off by this blanket of smoke; dirty curtains and decaying stonework. Then Professor Cohen made a perceptive suggestion, reported by *The Times* as follows:

> in the opinion of the Leeds [Smoke Abatement] Society, the Alkali Acts ought to be extended to cover all factory chimneys, and the control of smoke emission ought to be placed in the hands of the inspectors of the Local Government Board.[44]

Already in 1898 smoke-abatement societies in several northern towns had joined with Leeds to present to the Local Government Board a memorial on this matter. But owing to the apathy and indifference of the departmental officials the memorial had never been presented to the Board.[45]

One outcome of the 1905 conference was a consolidation, four years later, of smoke-abatement lobbies in the north, under the label Smoke Abatement League of Great Britain; to the embarrassment (incidentally) of the Coal Smoke Abatement Society, which did not want its campaign for London to be complicated by association with the pollution problems of the north.[46] Just as the London County Council was struggling to get its Bill through parliament, the Smoke Abatement League sent a deputation to the Board (on 29 June 1910) asking for an extension of the LCC Bill to cover the whole of the country, and pressing Professor Cohen's suggestion, focused as a proposal to set up a Smoke Department of the Local Government Board, with a staff of scientifically trained inspectors similar to those in the Alkali Inspectorate.[47] The President of the Board 'after hearing their views, expressed sympathy with their object and promised to do what he could, pointing out, however, the difficulties in the way'.[48] This of course was a diplomatic brush-off. The President fulfilled his promise by doing nothing.

The 'difficulties in the way' was a euphemistic expression for the mobilized opposition of industrial air polluters. The next step to ease these difficulties came not from the government but from the miners. On 18 January 1912 there was a two-thirds majority in a ballot to strike for a minimum wage-rate. Negotiations became deadlocked. By 1 March 850 000 miners came out on strike and by 10 March a further 1 300 000 workers were unemployed. The price of coal, and therefore economy in the use of coal, suddenly rose to top priority in the public mind. People began to ask what would happen when, one day, the coal seams gave out. *The Times Engineering Supplement* for 27 March reported a powerful plea for 'the most rigid economy in every branch of fuel consumption' especially as one benefit of economizing would be that the air could be 'purified from the blighting influence of smoke'.[49]

Arguments about conservation of coal supplies and economy in their use touched the hearts of two sections of British society which had hitherto given little support to the smoke-abatement lobby: industrialists and businessmen. The first two leading articles in a Fuel Number of *The Times* (on 1 December 1913) were 'Economy in Fuel' and 'Conservation of Coal Supplies'. To give edge to these themes there was a need for hard data. The CSAS began to collect it and *The Lancet* published the results: an estimated 76 000 tons of soot—something like 650 tons per square mile—fell on the administrative county of London each year.[50] Someone estimated that the economic cost in wasted fuel falling as soot over the whole of Britain represented the work of a million miners for three days! By 1914 fifteen more cities had agreed to collect similar data. This was the beginning of an annual monitoring of smoke, later conducted on a more sophisticated scale by the Department of Scientific and Industrial Research.

Statistical data on air pollution fortified the arguments of the smoke-abatement lobby in favour of tougher legislation. But there was still perplexity about how to deal with domestic smoke. Electricity was too costly for heating, though gas was becoming more competitive and was being vigorously marketed. The technological bottleneck was a supply of low temperature combustion coke. A patent for the production of coalite (by carbonization at about 600 °C instead of at 1200–1400 °C as used in the manufacture of coke) was lodged in 1906 by Thomas Parker. But the patent could not be commercially exploited without an assured demand, and the government was unwilling to create a demand by obliging people to burn smokeless fuels until there was an assured supply; an administrative log-jam which was not cleared even after the Clean Air Act was passed in 1956. But the delay in dealing with domestic smoke was not solely—or even mainly—due to lack of appropriate technology. There remained—there still remains—a deep 'sentimental attachment to the cheery glow of the old fashioned grate'.

Another attempt at legislation

The Smoke Abatement League, despite the failure of their deputation to the President of the Local Government Board, decided to press for a Bill to amend the Public Health Act of 1875 in order to bring smoke under a control as efficient as the control of noxious vapours. They were now in a stronger position, for they had the moral support of the electricity and gas industries and they could rely on a spreading concern for the economical use of coal. They secured the help of Gordon Harvey, MP for Rochdale and himself a cotton manufacturer. He offered to introduce a Bill in the House of Commons. It was duly introduced on 30 April 1913.[51] The Bill still fought shy of touching the domestic hearth but for all other fireplaces and furnaces it proposed to lift the perverse word 'black' out of the statute book and to deem smoke of any colour to be a potential nuisance; it proposed that the Local Government Board should have power to appoint smoke inspectors with powers analogous to those of the alkali inspectors; that local smoke-abatement authorities should be set up to administer the law; and that there should be more severe penalites. The Bill lapsed without a second reading and was not revived in the following session because its supporters had no luck in the ballot for private Bills. The House of Lords, however, rescued it: Lord Newton introduced a similar Bill there on 17 March 1914.[52]

Lord Newton's Bill was eclipsed by war and no further Bill appeared until 1922, but for three reasons it was an important event in the history of smoke abatement. First, the prototype of the Bill read in the House of Commons had actually been *drafted* by the Smoke Abatement League—a remarkable testimony to the influence of lobbies in the campaign.[53] Second, Newton rested his argument upon the conviction that public opinion had changed 'due more than to anything else to the invaluable work which has been performed by the various smoke abatement societies'. Public opinion was ripe, he said, for a change in the law.[54] Third, although the Bill lapsed, there was no opposition to it, and it was withdrawn on the announcement that the Local Government Board would set up a departmental committee to examine the whole problem of air pollution.[55] Thus began a convergence of the policies on noxious vapours and on smoke.

Meanwhile the Alkali Inspectorate, encouraged by the vote of confidence given to it by the Act of 1906, and now authorized to tackle the intransigent problems of smelting works, was able to work free from the tiresome political inertia which dogged the path of anti-smoke lobbies. In 1910 W.S. Curphey succeeded Carpenter as Chief Inspector. He was in the true apostolic succession, for he had joined the Inspectorate in 1877 as an assistant to Angus Smith and served in that capacity for five years before being promoted to a sub-inspectorship; thereafter he became an inspector for Scotland.

Curphey had only four years in office before the war disrupted his plans for the Inspectorate, but even in this brief period there were signs that he took a wider view of his remit than Carpenter had ventured to take. He gave a lot of thought to the abatement of gases from smelting works[56] and he stressed the value of mechanical appliances for controlling the processes which, in the past, had caused pollution because of inefficient manual operation: cement, sulphuric acid, chemical manure, and chlorine works.[57]

All this promising work had to be shelved during the war. The staff of the Inspectorate was depleted (by 1916 only 5 out of 9 district inspectors were still at their posts). Standards of abatement deteriorated and for a time no one was pulled up for infractions of the Alkali Act.[58] Nevertheless Curphey had time to think about the problems that would arise after the war, when industry reverted from making munitions to making products for peace. He foresaw that there would be severe industrial competition and that the chemical industry—enormously stimulated by the war—would have to be run on more scientific lines if it was to survive. Accordingly his reports carried exhortations to industry to manage its affairs more scientifically, and to employ more science graduates. It had seemingly needed the 'upheaval of war' (as one of his district inspectors put it) to convince manufacturers that it paid to employ university graduates in their works.[59]

8
CONVERGENCE

The Newton Report, 1921

Between 1914 and 1920 there was a watershed as steep as any in the social history of Britain. Yet despite the massive human problems of post-war reconstruction, which might have been expected to leave no time for concern about the environment, there was a prompt rekindling of interest in smoke abatement. Less than four months after the armistice Sir William Richmond was before the public again, writing in *The Times* about the way coal was being wasted as smoke.[1] Ten months later in January 1920, the departmental committee announced by the President of the Local Government Board in 1914, was reconstituted (this time by the Minister of Health, whose department had taken over the Board's responsibility for abating air pollution), with Lord Newton as chairman.

This is just what Lord Newton had asked for when he presented his Bill in 1914. He knew (and said in his introductory speech at the time) that there was not the remotest chance of his Bill becoming law; all he wanted was consent from the government to have an inquiry into the state of the law on all kinds of air pollution.[2] As in 1914, the inquiry was to cover noxious gases as well as smoke, 'and to advise what steps are desirable and practicable with a view to diminishing the evils still arising from such pollution'. The very wording denoted a change in public attitude. No longer could industrialists expect to defend smoke simply by saying that where there's muck there's money. Pollution was an 'evil' and the onus was now on the polluter to justify his pollution or to get rid of it. The Committee (which included the Chief Alkali Inspector, Curphey, although he was just about to retire) set to work with great energy. Within two years, after nearly 50 meetings and the examination of 150 witnesses, it produced a report which was a model of brevity and lucidity.[3]

A brief and lucid report; but that was about all that could be said for it. In its disarming lucidity it failed to confront the complexity of the issues. Although the need for a smoke inspectorate organized centrally on the model of the Alkali Inspectorate was advocated in the evidence, the Committee rejected this and made no attempt to disturb the two-tier system for the control of air pollution: noxious gases were to remain the responsibility of a team of scientists under the Ministry of Health; smoke and other unregistered pollutants were to remain the responsibility of local authorities. The gist of the recommendations was to endorse the existing framework for control and to reinforce it.

The Alkali Act should be reinforced, said the report, by putting into it a list of noxious gases and giving the Minister power to add to the list by Order and to register under the Act any process or works which emitted any of these gases (the very principle which Fletcher tried in vain to have put into the amending Act of 1892). And 'every manufacturer should be put under the general obligation of using the best practicable means for preventing the escape of noxious or offensive gases, and ... no limiting figure should be laid down by the Act as to certain gases'. The point of this last recommendation was that fixed emission limits deter manufacturers from improving their techniques for abating pollution and offer no spur to further research. It would be sufficient, said the report, for the Minister, after due consultation, of course, to set 'presumptive standards' representing the level of abatement which was reasonable in the current state-of-the-art.[4]

Over smoke the report had more substantial proposals. The first was to get rid of the exasperating word 'black' and to oblige manufacturers to use the best practicable means (a testimony to the success of this formula in the control of noxious gases) to abate smoke of any colour; with a rider that the Ministry should have authority to issue presumptive standards for smoke if this proved advisable. The Committee did not face up to the fact that standards for smoke—undoubtedly advisable—were not technically practicable.[5] A second proposal was that responsibility for enforcing that the best practicable means were being used should no longer rest with the 1800 sanitary authorities, but should be transferred to the fewer and more affluent county authorities, which could be expected to employ better trained and better paid inspectors.[6] A third proposal was that the Ministry of Health, though it should not take control of smoke abatement, should nevertheless have emergency powers to compel slack local authorities to do their duty. The only concession to the pressure to set up an inspectorate on the model of the Alkali Inspectorate was a recommendation that the Ministry should provide some sort of advisory service to help local authorities over their smoke-problems.[7]

So confident was the Committee that domestic smoke could be controlled that it urged the government, in a brief interim report soon after it was appointed, to make it obligatory for government-subsidized housing to be equipped to burn smokeless fuel.[8] It was a critical moment for this, because the building of homes had a very high priority in the programme of post-war reconstruction. It could have been done. The old fashioned kitchen range was already on the way out and heating by closed stoves or gas had been available at reasonable cost since the end of the nineteenth century. But no politician was prepared to deprive the Englishman of his companionable, pokeable fire. On this social inertia that blocked the way to smokeless heating the report had hard things to say. Two members of the Committee went to the industrial area of Rhine-Westphalia to see how the

Germans abated smoke. German homes, they found, were 'practically smokeless'; they were heated by closed stoves burning coke or briquettes. There was a well enforced control of industrial furnaces, which were not permitted to burn bituminous coal. As as result great towns like Dusseldorf were pleasant and agreeable places of residence.[9]

The report came at an unpropitious time. In 1921 the post-war boom collapsed, exports fell, unemployment soared, the owners of coal mines cut wages and met union resistance with a lock-out. The one comfortless benefit from this crisis was the sight of London as a 'smokeless city' and the reminder that to waste coal was to endanger Britain's economy. When in 1922 a meeting was held in the Ministry of Health to discuss Newton's report, the chairman, an assistant secretary in the Ministry, 'laid it down definitely that legislation on the report was out of the question at present, particularly any legislation which would involve the creation of additional officials'.[10]

This was the advice the civil servants gave to the Minister. But the Minister, appointed by Lloyd George in 1921, was the industrialist Alfred Mond. He was the first minister to display an active interest in smoke abatement since Palmerston's tenure at the Home Office in the 1850s. He rejected the official line given to him in his brief, and assured a deputation of the Coal Smoke Abatement Society, when they came to see him about the report, that he would take the report seriously.[11]

He was as good as his word. On 10 May 1922 Newton's report was debated in the House of Lords. The government spokesman (the Earl of Onslow) announced that action would be taken. Before the government introduced a Bill, he went on to say, 'We are awaiting the receipt of the draft Bill from the Coal Smoke Abatement Society'.[12]Here was a triumph for the smoke-abatement lobby: for a second time they were invited to submit their own draft as a basis for legislation.

On 19 July the Earl of Onslow did introduce the Bill, too late in the parliamentary session for it to have any prospect of survival; but that did not matter: its purpose was to test the temperature of public opinion. Newton was disappointed, for the Bill went only a little way toward the (admittedly impracticable) reforms his Committee had proposed.[13] It was, he said, a 'weak and faltering step' toward the aspirations of his Committee; he felt 'in the position of a man who has been promised a passage, say, to America, and has been given only enough money to take him to Southampton or Liverpool.' Over noxious gases the government refused to depart from its long-standing custom of piecemeal additions to the powers of the Alkali Act; they would not have an umbrella clause covering all emissions of listed noxious gases or a comprehensive obligation put upon manufacturers to use the best practicable means to abate pollution. Over smoke also the Bill fell short of the Committee's recommendations: it made no mention of compliance with presumptive standards.

'Weak and faltering' the Bill may have been, but it was a sufficient signal for the opponents of smoke abatement, hitherto silent, to declare themselves. In March 1923 a pressure group from the Federation of British Industries waited upon the Parliamentary Secretary to the Ministry of Health, who had introduced the Bill.[14] Their case was put by a tetchy and combative academic, Professor Bone, who held the chair of fuel technology at the Royal College of Science. He spoke of misconceived revision of the law that might 'imperil the industry in this country and absolutely ruin the commercial efficiency of trade'. He adopted the familiar tactic for stalling. What was wanted, he said, was a much more thorough survey of the problem.

The Public Health (Smoke Abatement) Act, 1926.

The history of the campaign for clean air is littered with abandoned Bills. We have already recorded examples and we shall not weary the reader with a recital of the Bills based on Newton's report which sank without trace between 1922 and 1925. Finally, in 1926, the seed sown originally in 1914 came to harvest. On 11 March 1926 there was launched in the House of Lords the government Bill which put some of the recommendations of Newton's committee on the statute book. It was called the Public Health (Smoke Abatement) Bill.[15] The clauses of the Bill referring to noxious gases passed through parliament with comparatively little discussion. There were two useful provisions. One was to give the alkali inspectors (now under the Ministry of Health) authority to inspect unregistered works if they were deemed 'likely to cause the evolution of any noxious or offensive gas'. This brought within the purview of the inspectors all sorts of processes which had hitherto been out of their reach. The other useful provision was to simplify the procedure for registering new gases as noxious and new works to be brought under the Act. Hitherto all additions to the list of registered works had required the introduction of formal legislation. Under the 1926 Act all that was required was a Statutory Instrument to be laid before both houses after prior consultation with 'interested parties'. If no objection was raised to the Order in the ensuing 28 days it became law. By this means 17 new works and some 224 new processes were promptly brought under the control of the Alkali Inspectorate.[16]

The clauses about smoke in the Bill had a more difficult passage. In the House of Lords there was outspoken criticism of the government's capitulation to shipping interests—steamships were to be exempted from control—and there was a tough effort to bring domestic smoke under restraint.[17] In the House of Commons there was an amendment (which was defeated) to restore the old qualification of 'black'[18] and renewed pressure there also to deal with domestic smoke. Sir Alfred Mond urged the House to encourage the trend, already evident, to burn smokeless fuel in domestic grates.[19] But in vain; there was the inevitable genuflection to the Englishman's open fire, and a rearguard action against the use of gas which deserves a footnote in the history of clean air. It came from Mr Storry Deans. 'Although I may be told', he said, 'that the smoke from my coal fire assists in poisoning the people outside, I prefer that very much to being poisoned myself by a gas fire within my own house.'[20]

The issue was pressed to a division and an amendment to end the privileged status of the domestic stove was lost by 192 votes to 54.[21]

After all the tinkering with the drafts the smoke clauses in the Bill emerged in such an enfeebled state that they were no more than trivial amendments to the Public Health Act of 1875. The penalties for pollution were slightly stiffer: £50 instead of £5 as the maximum fine—not exactly an ominous threat to a large industry. The definition of smoke was extended to include soot, ash, and gritty particles. Local authorities were given legal powers to set standards for abatement. The central government had slightly more powers to intervene and to assist in the administration of the law by local authorities. But there was no release from the crippling qualifications in the Public Health Act 1875 or in the Act of 1891 applying to London. Instead there was a fresh confusion, for there was a sub-section which seemed at first sight to get rid of the necessity to prove that smoke was black. It ran as follows:

> For the purposes of section 91 of the Act [the Public Health Act 1875] a chimney (not being the chimney of a private dwelling house) sending forth smoke in such quantity as to be a nuisance shall be deemed to be a nuisance...notwithstanding that the smoke is not black smoke.

But the Act goes on to say:

> In any proceedings for sending forth smoke, *other than black smoke* [our italics] from a chimney in such quantity as to be a nuisance, it shall be a defence for the person charged to show that he has used the best practicable means for preventing the nuisance.

So: if your smoke was not black, you were protected by the defence that you had used the best practicable means; if your smoke was black, this defence would not be accepted, but the court had to be satisfied that the smoke really was black. The one improvement was that the onus now lay with the defendant to prove that he had used the best practicable means. The problem of defining black smoke remained to confuse the courts. No progress had been made in the definition of non-black smoke 'deemed to be a nuisance'. And the formula 'best practicable means', so simple to define for the control of hydrochloric acid—and therefore so effective in the hands of the alkali inspectors—remained extremely difficult to define for smoke.

Public opinion matures

It is no wonder that the control of noxious vapours and of smoke have had such separate and contrasting histories. Even to this day, there is no satisfactory technique for measuring the density of smoke. The length to which people have gone to devise techniques is exemplified by an attempt made in the 1950s. A unit of smoke, felicitously called a 'murk', was suggested. A murk was defined as the amount of dirt in the air which produced a decrease of 0.1 in the logarithm of the reflection factor on one square millimetre of filter paper through which 0.1 cubic metre of air had been drawn. This monstrosity illustrates the difficulty. When, in 1956, a standard was ultimately issued, it was a much cruder subjective test. The density of smoke was compared by eye with the density of lines on a sheet of lattices called a Ringelmann chart, illustrated in Fig.2 (p. 108).

The measurement of smoke, like the measurement of noise—an equally frustrating problem—involves several dimensions. Is it intermittent or persistent? (Intermittent smoke is unavoidable from a great many sources.) Is it offensive in colour? Are its effects exacerbated by the presence of sulphur, grit, and dust? Faced with such complexities as these, the politician despairs of controlling smoke (and the same is true of noise) by mere legislation. The only hope for effective control is to secure a consensus in the population that smoke is intolerable. This is what the sanitary reformer, John Simon, urged so forcibly a hundred years ago in the passage we quoted on p. 54.

So the labours of the smoke-abatement lobbies were by no means over. In 1929 the Coal Smoke Abatement Society and the Smoke Abatement League coalesced to form the National Smoke Abatement Society which, as the National Society for Clean Air, is still the recognized and respected guardian of clean air in Britain. It soon became apparent that local authorities were shy of making much use of their mildly greater powers under the 1926 Act. By 1932, some 155 authorities had framed by-laws to fix local standards of emission, but in every case these referred only to black smoke and its duration.[22] A questionnaire issued by the Ministry of Health to a selection of local authorities in that same year elicited the significant information that virtually no proceedings had been taken against emissions of non-black smoke.[23]

All the same, despite the glacial progress of anti-smoke legislation, public opinion—the most powerful force for reform—was on the move. On every side there were signs that the pacemakers for clean air were drawing support. Industry was beginning to make a constructive response. In 1937, for instance, the world's largest low temperature carbonization plant was opened at Bolsover in Derbyshire. By 1938 less than half the energy for domestic consumption came from bituminous coal and 60 per cent of the output of gas was for domestic purposes. And—strongest evidence of all—

in the 1930s the coal lobby, a Goliath among pressure groups, was becoming uneasy about the tide of enthusiasm being generated by the David of pressure groups, the National Smoke Abatement Society. In 1935 the Coal Utilization Council held a convention to boost the use of bituminous coal. Speakers derided the smoke-abatement lobby as a bunch of idealists with woolly theories and hinted that they were financed by vested interests in gas, electricity, and smokeless fuel. John Charrington—a major coal merchant—warned his colleagues against the growth of public opinion and the possible future trend of legislation and he catalogued the virtues of the coal fire, including, of course, its 'friendliness', and adding two novel qualities: the 'physiologically stimulating' variations in temperature in the room and the subtle yet potent 'biovitric rays' which, according to one enthusiast, suffused from the companionable open hearth.*

After the Second World War the movement toward smokeless fuel gathered momentum. An Advisory Council of fuel and power under the chairmanship of Sir Ernest Simon and with a membership of unmatched distinction (it included two future presidents of the Royal Society) produced a report in 1946 which demolished all rational arguments against the phasing out of raw coal in domestic houses, leaving opponents no defences except prejudice and sentiment.[24] On every other count—health, cost, efficiency, and convenience—smokeless fuels won the argument. The only concession to traditionalists was that the transition would take 20 to 30 years before a sufficient supply of smokeless fuels could be assured and before the nascent social values for clean air would become the normal expectations of the British public. But in an appendix to the report there was a sardonic warning: the transition could not be expected to take place without positive leadership from the government:

> Smoke abatement, like any other measure of public health, must come, not from the free choice of individuals, but from concerted action by the State through the agency of local authorities.[25]

The Advisory Council's judgement was at fault here, for although the State in a pluralistic democracy can urge, it cannot, in an issue like this one, compel. Shortly after the Advisory Council's report was issued there was a notable example of the power of the 'free choice of individuals' in the city of Coventry. The city authorities drafted a Bill to give themselves power to create smokeless zones. It was opposed. The authorities countered by holding a referendum. Although the conversion to smokeless stoves would

*The enthusiast was Dr Marie Stopes, a distinguished coal-geologist, better known as a pioneer in the advocacy of birth control. Although her knowledge of coal-geology was sound and her views on contraception very courageous—if sometimes sentimental—she did harbour some crazy notions, one of which was the 'vital radiance' emanating from open coal fires. What the open coal fire undoubtedly does do, and central heating does not, is draw air up the chimney and so ventilate the room.

entail expense, the citizens of Coventry voted in favour of smokeless zones by 27 990 votes to 11 302.[26] Manchester had already acquired powers to create smokeless zones in 1946; Coventry obtained these powers in 1948, and similar enactments were made for other local authorities. Nevertheless the government still dragged its feet over the extension of powers to cover domestic smoke. The reason was that it believed it prudent to delay any revision of the law until there were assured supplies of the sort of smokeless fuel that could be burned in an open grate.

Smoke and the alkali inspectors

Throughout this episode in the history of smoke abatement the Alkali Inspectorate became more and more closely involved, though always informally and never with specific administrative responsibility. A Chief Inspector with the energy and enthusiasm of Fletcher would surely have seized the opportunity offered by the report of Lord Newton's committee in 1921 to involve the Inspectorate in smoke control. Curphey, too, who was Chief Inspector from 1910 to 1920 and was a member of Newton's committee, might have brought policies for the abatement of noxious gases and smoke closer together. But Curphey retired in 1920, and his successor, T.L. Bailey (who served as Chief Inspector from 1920 to 1929), clearly did not want to be bothered with smoke. He was asked by the Ministry of Health (the Ministry under which he served) to help over the devising of presumptive standards for the emission of smoke and the Ministry turned to his Inspectorate to provide the 'one or more competent officers' to advise local authorities and manufacturers on difficult smoke problems recommended by the Newton Report.[27] Over neither of these matters was Bailey co-operative. He was, of course, quite right to discourage the Ministry about presumptive standards for smoke; he knew just how difficult it was to describe intensities of smoke. He was, perhaps, less right to turn away the opportunity to involve his staff in advisory work on smoke. He flatly 'considered that the present staff of Alkali Inspectors could not undertake any additional duties'.*[28] He was, it is true, understaffed and new duties were being imposed on his Inspectorate owing to the growth of the chemical industry. But basically Bailey, as one of his successors wrote, 'was a bit lukewarm on smoke. His successor and my predecessor—Mr W.A. Damon,

*He was successful in protecting his staff from additional duties. But the Ministry of Health, without waiting for the passage of the 1926 Bill, agreed to provide advice to local authorities. The job was given to three officials of the Ministry: Bailey himself, H.F. Hooper from the Engineering Inspectorate, and J.C. Dawes from the Public Cleansing and Salvage Inspectorate; but it soon devolved on Dawes who already had experience of smoke abatement with local authorities. He became the expert on smoke abatement. Bailey, apparently, opted out (HLG 55/17; 55/8).

1930-1955—became interested in the mid-1930s... and later violently concerned and pulled a faintly protesting inspectorate along with him'. To which the writer (J.S. Carter, see pp. 120-4) added as an afterthought: 'but without statutory authority and using the alkali inspectors' powers of entry under the Act of 1926.'[29]

So the Act of 1926, weak and permissive as its terms were, did enable the Alkali Inspectorate to extend its beneficent influence over smoke, under a Chief Inspector who reigned for 26 years. To his distinguished tenure of the office we now turn.

His name was W.A. Damon. He had worked for eleven years in industry before joining the Inspectorate in 1921. He was chosen to succeed Bailey in December 1929. He was retained for four years by the Ministry of Health after his retirement from the chief inspectorship in 1955; so it was not until 1959, after 38 years in government service, that he finally retired. 'Few men', it was said of him at his retirement, 'have worked harder for the cause of clean air.'[30]

Damon came into office at the time of the depression. Any prospect of establishing a smoke inspectorate was out of the question on financial grounds and the faithful J.C. Dawes was putting aside what time he could spare from cleansing and salvage work to advise local authorities and the regional committees set up according to a suggestion made in the Newton Report in 1921. The technical work was not being co-ordinated in any way (complained Dawes) 'and never has been although it has mostly fallen to me'.[31] Dawes was still pressing in 1932 for the appointment of someone in the Ministry of Health who would give the whole of his time to advisory work on smoke abatement. A minute written by one of the Ministry's officials (Ross), commenting on Dawes's request, shows how civil servants, far from ignoring the 'free choice of individuals' (as the Ministry of Fuel and Power had been advised to do)[32] were waiting for a lead from public opinion. Ross wrote:

> It seems unlikely that any big drive in smoke abatement will take place until a definite relation of cause and effect has been established (it has not) between atmospheric pollution and injury to health...[33]

(It is a pity Ross did not know his history better; if he had, he could have reminded his Minister that John Graunt in 1662 and Rollo Russell in 1880 had at any rate produced evidence of a correlation between air pollution and health;[34] and it was in fact just such another correlation—deaths caused by the London smog of 1952—that precipitated the Clean Air Act twenty years later.)

When Bailey was Chief Inspector, Dawes got little help from him. But under Damon the Alkali Inspectorate began to take a keen interest in smoke. Every annual report from the Chief Inspector included some discussion of the problems of smoke abatement; and in 1936 a new administrative arrangement was made whereby the Alkali Inspectorate became charged with the duty of advising the Minister on developments in smoke abatement.[35] It was one more step in the slow progress toward a consolidated clean air inspectorate. In 1936 representatives from five out of the seven regional committees on smoke abatement attended the annual staff meeting of alkali inspectors. Damon's annual reports carried blunt criticism of the lack of progress in smoke abatement. Much air pollution from industry was still due to carelessness in stoking, insufficient boiler capacity, and bad draught regulation: precisely the faults diagnosed by Mackinnon nearly a century earlier, and all curable.[36] Damon's reports exposed also some ingredients of coal smoke which were not so easily removed as the smoke itself could be, namely grit and dust and sulphur dioxide. It is interesting that as long ago as 1886 the annual report on the Alkali Inspectorate mentioned experiments by the physicist Oliver Lodge (1851–1940) on electrostatic precipitation of grit and dust, but it was not until the 1920s that this technique was applied to the removal of pollutants on an industrial scale. Its first application to boiler flue gases was at the Willesden Power Station in 1929. As for sulphur dioxide, Damon deplored (as long ago as 1936) the 'lamentable paucity of data' on techniques for removing sulphur from coal.[37] This is a problem for which there is still no economically feasible solution.

By the time the Second World War came to interrupt, a second time, the progress of the Alkali Inspectorate, Damon had firmly established that his extrastatutory duties—the surveillance of smoke—were as important for his Inspectorate as were his statutory duties. Early in the war the production of dark coloured smoke was positively encouraged as an additional protection against enemy attack from the air; though this order was waived in 1942. The work of the regional smoke committees, never very effective, was suspended. But as soon as the war was over Damon resumed his interest in smoke abatement. Gradually he contrived to blur the distinction between this non-statutory work of the Inspectorate and their statutory work under the Act of 1926, of inspecting premises likely to cause the evolution of noxious or offensive gases. By 1948 some 30 per cent of the Inspectorate's time was spent on problems at these various unregistered works,[38] where (as Damon emphasized in his annual reports) the responsibility for taking action rested with the local authorities; all the alkali inspectors could do was to offer advice.

In interpreting his duties to cover 'the reduction of atmospheric pollution by industrial processes' (this was his declaration of purpose in his annual report for 1946) Damon evidently had the support of his Ministry[39] and he had the support, too, of industry. The journal *Chemistry & Industry* wrote:

> It is very good to see to what a large and increasing extent the services of the inspectors are used by various Ministries, Government departments, local authorities and other bodies in connection with the safeguarding of public amenities.[40]

Damon's determination to involve his Inspectorate in all kinds of industrial air pollution put a strain upon the small staff, and to some extent diluted their attentions to the processes over which they had statutory authority. But there is no evidence that this led to any laxity in surveillance. The Inspectorate followed its traditional style. Works emitting noxious gases remained unregistered—though they could be visited and the emissions studied—until there was some prospect that the gases could be abated by some best practicable means. When, and only when, something definite could be done to abate the gases, the Ministry of Health was asked to issue Orders for the works to be put on the register. This was done in 1935, 1939, and again, after the war, in 1950. The orders covered gases from artificial silk manufacture, the refining of paraffin oil, lead works and works emitting fumes containing cadmium; also fluorine, especially from brick kilns, where it had been shown to harm the health of cattle. Parliament let through these Orders without comment, but this is not to say that the issue of an Order is a formality, for in 1949 it was proposed to add carbon black to the list of registered processes. This was opposed when circulated in draft-form before reaching parliament, and carbon black was left out of the list.[41]

At mid-century, in 1950, the Alkali Inspectorate was still essentially the same tiny group of scientists as it had been at the time of Lord Newton's inquiry thirty years earlier.* Its style of working was in the tradition set by Angus Smith: pragmatic, flexible, forbearing in difficult cases, strict where strictness was justified. Thus, over cement works—a chronically tiresome topic—the line taken was that each works had to be considered separately; hard-and-fast rules were not laid down; the size of the works, the local conditions, the age of the equipment: all these had to be taken into account.[46] So also with sulphuric acid works. 'District inspectors', wrote Damon, 'have...been instructed to regard as infractions of the Act only such escapes at old works as correspond to more than 5% of the sulphur burnt' (the requirement for new works was 2 per cent).[47] But in his first post-war report, for 1939-45, Damon gave notice of his intention to press for more stringent control in sulphuric and nitric acid works, and he warned that war-time indulgence was coming to an end. Throughout the whole of his twenty-six years in office, Damon brought only two prosecutions. His weapon, save in cases of wilful or flagrant transgression, was a written complaint to the offender and a recorded list of infractions published in his annual reports. For works subject to the requirement of best practicable means, he wrote: 'There is no well defined dividing line between legal and

*In 1920 there was a Chief Inspector and 7 district inspectors for England, Wales, and Ireland (the grade of sub-inspector was dropped); also (created in that year) a Chief Inspector for Scotland.[42] By 1950 the strength of the main Inspectorate had been augmented by a clerk[43] and the rank of deputy chief inspector was created[44] but only one inspectorship had been added to the establishment.[45]

illegal operation and a decision as to whether a given set of conditions shall or shall not be treated as constituting an infraction rests, to a great extent, with the District Inspector.'[48]

The remarkable thing about this velvet glove technique was that it was really delegating to the inspectors quite difficult quasi-political decisions. Every decision the inspector made could, by accepted usage, take into account not only the technical feasibility of abating the noxious gas, but also the finances of the firm, the age of the equipment, the morale of the operatives and the effect on their employment, the national need for the product, and the amenity of the public. Yet there was in the 1950s practically no parliamentary criticism of the Inspectorate and there was strong support from industry. The Inspectorate had a low profile among the general public but it was recognized, in Britain and abroad, as a model scientific service for the protection of the environment. The journal *Chemistry & Industry* in its regular reviews of the Inspectorate's annual reports, reflected the attitude of those who were most closely in touch with the problems:

> The alkali inspectors are a remarkable body of men. Any inspectors who are not looked upon by the inspected as having only nuisance value must be remarkable, and that is most certainly the case with the alkali inspectors. They have to administer an Act and a number of Orders which could easily become an intolerable nuisance to industry were they not intelligently and helpfully administered.[49]

Much of the credit for this compliment must go to Damon's leadership. In another review of the Inspectorate's work, *Chemistry & Industry* wrote:

> The avuncular activities for which Mr. Damon was so well known before the war are again obvious in this Report, but Dr. Balfour Birse [Chief Inspector for Scotland] ...runs him a close second. But these uncles can be stern on occasion as may be read on more than one page...[50]

If the control of noxious gases was—with some awkward exceptions—reasonably satisfactory, the same could not be said about the control of smoke, grit, and dust. The reasons for this contrast were clear. Gases could be monitored quantitatively, smoke could not. Gases were under surveillance by a team of experts, the surveillance of smoke was left to local public health officials who had many other duties and most of whom lacked the technical knowledge to deal with it. Gases scheduled under the Alkali Act were identifiable as the waste products from comparatively few industrial processes, smoke poured from millions of homes. No industrialist would claim that his noxious vapours were beneficial; millions of citizens cherished the benefits of an open coal fire and were willing to put up with the inconvenience of smoke. Some sort of shock was needed to galvanize public opinion. The shock came in December 1952.

9

BREAKTHROUGH FOR SMOKE ABATEMENT

The London smog of 1952

On Friday 5 December 1952, preparations were under way for the Smithfield Show of livestock at Earls Court in London, which was to open on the following Monday. It was on that Friday night that an unusually nasty fog descended on the city. Londoners were used to fogs. Their reaction to it was described afterwards as 'strangely calm...almost fatalistic'.[1] On Monday the fog was as thick as ever. *The Times* reported that some of the cattle brought to Earls Court were suffering from respiratory trouble.[2] An Aberdeen Angus died; 12 others had to be slaughtered; and about 160 of them had to be treated by vets.[3]

These were the first reported casualties in the smog of 1952. There were other unusual incidents. At Sadlers Wells theatre the opera *La Traviata* had to be abandoned after the first act because the audience could not see the stage.[4] By Wednesday the fog had cleared, but its effects were only just about to begin. In the following week the Minister of Health was asked in the House of Commons how many additional deaths had occurred in the metropolitan area. During the week ending 13 December, he replied, there had been 4703 deaths compared with 1852 deaths in the corresponding week in 1951.[5] At the end of January 1953, some six weeks after the fog, the London County Council issued a dramatic report on the fog. It gave figures for the excess of deaths per million inhabitants over the normal number in the LCC area attributable to four disasters between 1866 and 1952.[6] The figures are shown in Table 1.

Table 1

Date: week ending	4.8.1866	20.12.1873	19.11.1918	13.12.1952
Cause	cholera	fog	influenza	fog
Total deaths	876	713	1085	745
Normal number of deaths	450	470	300	300
Excess deaths over normal	426	243	785	445

The ultimate number of deaths attributable to the 1952 smog was, of course, much greater (a total of some 4000), for many people who survived the week ending 13 December died later on from circulatory or respiratory disorders brought on by the smog. The grim lesson from these figures was that fog in the mid-twentieth century was as dangerous a killer as cholera had been in the mid-nineteenth century.

This information was not new: Rollo Russell published data just as grim as far back as 1880.[7] But this time, in 1952, the incident crystallized public opinion in the perplexing way such incidents do. It is true that there were better and cheaper means for abating smoke in 1952 than there were in 1880, but it would be a gross over-simplification to attribute the impact of the 1952 smog, compared with the lack of impact of scores of earlier fogs, to a change in technical and economic circumstances alone. The overriding circumstance was the ripening of public opinion.

This became evident on 20 January 1953, when parliament reassembled after the Christmas recess. Day after day members tackled ministers at question time. The ministers, unprepared for this indignant demand for action, prevaricated. 'I am not satisfied', said Mr Macmillan, speaking for the Ministry of Housing and Local Government, 'that further general legislation is needed at present'. Pressed to urge local authorities to exercise the powers over smoke they already possessed under the Public Health Act 1936, he replied: 'We do what we can but, of course, the hon. Gentleman must realise the enormous number of broad economic considerations which have to be taken into account.'[8] The spearhead of criticism in parliament came from Dr Dodds and Lt.Col. Lipton. By the end of January they were accusing the government of complacency and 'an amazing display of apathy'.[9] Throughout February the pressure on reluctant ministers continued. Some disquieting facts were squeezed out of them. Twelve local authorities, for instance, had powers by local acts to establish smokeless zones but only two of them (Coventry and Manchester)[10] had used their powers.[11] One reason for this—and it was a reason behind Mr Macmillan's evasive remark—was an insufficient supply of smokeless fuel. For the same reason the government dragged its feet over permitting the City of London to legislate against smoke. But it must not be inferred from this that the technology was not available, what was not available was the political will to apply it. The press kept the pot boiling too. The Minister for Health, Mr Macleod, had already declined to put a representative of his department on the Atmospheric Pollution Research Committee, on the ground that his department was concerned with the effects, not the causes, of fogs. When baited on his apparent lack of interest he was goaded to reply: 'Really, you know, anyone would think fog had only started in London since I became Minister'.[12]

There was no sign that the government was leading public opinion over this episode. On the contrary, it was dragged by criticism, and with obvious reluctance, to promise—but not till the following May—that a committee of inquiry would be set up. It was not until July 1953 that the government announced that a chairman had been appointed.[13]

The Beaver Committee

There was no excuse for this dilatory attitude on the part of the government. Four years before the London smog, in October 1948, there had been a similar disaster in Donora, an industrial town in Pennsylvania. A sulphurous fog descended on the town; nearly 43 per cent of the population fell ill from respiratory troubles and 20 persons died. The event shocked the American public. The Public Health Service sent a team to study the disaster and to recommend action. All this was known to the British government, for the Chief Alkali Inspector, Damon, had attended a conference on atmospheric pollution held in Washington in May 1950 and had published in his report for 1950 an account of the ordinances for abating smoke already operating successfully in Pittsburgh and St Louis. Damon was known to be holding the hand of his Minister (Duncan Sandys, Minister of Housing and Local Government) in the fuss that followed the London smog. He certainly gave advice on the composition of the Committee which his Minister set up in July 1953,[14] and—when the records become available—we expect to find that his advice was taken.* It is significant that one member (O.G. Sutton) and one assessor (A. Parker) were his companions on his visit to Washington to the air pollution conference; P. Lessing, a consulting chemist, is known to have been in close contact with Damon; and the members from industry—G. Nonhebel, Sir Roger Duncalfe, and the Committee's chairman, Sir Hugh Beaver, would all have had dealings with the alkali inspectors; and Damon himself was an assessor on the Committee. It is a safe assumption, therefore, that the traditions and style of the Alkali Inspectorate influenced the deliberations of the Committee.

The Committee's terms of reference were:

> To examine the nature, causes and effects of air pollution and the efficacy of present preventive measures; to consider what further preventive measures are practicable; and to make recommendations.

The government was by now so sensitized to criticism that it wanted a quick answer. Meanwhile the stock reply to parliamentary questions about air pollution was: we must await the report of the Beaver Committee.

There was not long to wait. Four months after it was appointed the Committee issued an interim report and its final report followed in November 1954.[15] The final report received an accolade of praise from press and parliament, and the government promptly announced its approval of the recommendations in principle (a formula which allows plenty of room for manoeuvre).

*Members of the Committee were: Sir Hugh Beaver (Chairman), Miss A.D. Boyd, Dr J.L. Burn, Mr S.R. Dennison, Sir Roger Duncalfe, Professor T. Ferguson, Dr G.E. Foxwell, Dr R. Lessing, Mr G. Nonhebel, Mr C.J. Regan, Professor O.G. Sutton.

There was nothing very novel or even original about the recommendations. Like most committees appointed to propitiate public demand, this one became a point of crystallization for reforms which had been adumbrated for years. Its claim to distinction was that it cleared the log-jam of obsolete laws and administrative practices which were holding up the free flow of technology into policies for clean air, and it put in their place proposals which made sense both to politicians and to administrators. The basic recommendation was that there should be a national Clean Air Act to replace all existing statutory provisions to abate smoke and that for the first time it should apply to smoke from domestic as well as from industrial premises. More important, perhaps, than this recommendation was the forthright assertion:

> we wish to state our emphatic belief that air pollution on the scale with which we are familiar in this country today is a social and economic evil which should no longer be tolerated, and that it needs to be combated with the same conviction and energy as were applied one hundred years ago in securing pure water. We are convinced that given the will it can be prevented. To do this will require a national effort and will entail costs and sacrifices.[16]

The assertion swept away the excuses which had been marshalled to postpone the campaign against smoke: that more research was needed before changing the law; that smokeless fuels were not sufficiently well developed; that they cost too much; that the public did not like them. It left only one explanation for any further delay: a lack of political will in Whitehall to carry out a reform for which people were now ready to pay. This was something the government could not afford to admit. Of course much more research and development were needed to solve many problems of air pollution. But the limiting factor which the Beaver Committee underlined was a lack of determination to make full use of technology already available.

The principal proposals in the Report were these:

1. Subject to certain exceptions the emission of dark smoke from any chimney should be prohibited. This simple prohibition would get rid of the tiresome need to prove that smoke was a nuisance. But what were the 'certain exceptions'? And what was 'dark smoke'? The Committee tackled both these elusive questions. There were three categories of exceptions. First, all furnaces are liable to smoke for a short time after they have been started up; this would have to be permitted. Second, some furnaces would not be able to satisfy a prohibition until they had been modified or replaced; so exemptions would have to be issued until modifications had been made. Third, exceptions would have to be made for emissions for which no practicable means had as yet been found to suppress the smoke. As for the word 'dark', the Committee adopted the standard used in the ordinance which had brought clean air to the city of Pittsburgh, namely that smoke is

dark if it has a density equivalent to, or greater than, shade 2 on the Ringelmann Chart. The chart (Fig. 2) is held at arm's length and the density of smoke is matched by eye against the four shades; a hazardously subjective criterion, and one which explains why it took so much longer to legislate for the abatement of smoke than it took for the abatement of noxious gases which could be accurately measured.

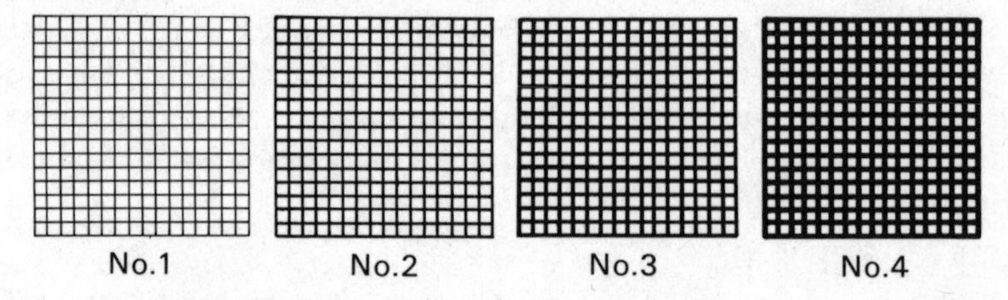

Fig. 2. The Ringelmann Chart. The chart is held at arm's length and the density of smoke is matched by eye against the four shades. The method of use is laid down in British Standard 2742:1969. 1 = light grey, 2 = dark grey, 3 = very dark grey, 4 = black. The Clean Air Act forbids the emission of smoke darker than Ringelmann 2, except for brief periods when the furnace is being lighted up or serviced.

2. Equipment for arresting grit and dust must be installed in any industrial plant burning solid fuel at the rate of 10 tons or more per hour. Note the reasonableness of the requirement: it was not to apply to furnaces where the cost might not justify the benefit. This was an endorsement of the emphasis which Damon had put upon the need to control grit and dust in all his annual reports since 1946.

3. Local authorities should have power, subject to the approval of the Minister, to issue orders creating smokeless zones and smoke-control areas. This was to replace the clumsy and more costly procedure whereby local authorities had to promote an act of parliament to get such powers. Credit for the idea of smokeless zones belongs to the National Smoke Abatement Society (who adopted it as a policy as long ago as 1936, and whose leader Charles Gandy was putting forward, even during the blitz in London in 1940, the plea that when the time came for rebuilding the bombed cities, it would be a wonderful idea to create smokeless zones in them). The decision to leave the initiative to the local authority was a realistic one, for it recognized respect for the autonomy of local government (always a delicate matter when laws to cover the whole nation are under discussion), and it ensured that a local authority would not take the initiative unless there was sufficient backing from the ratepayers to justify its action. To this recommendation the Committee added a politically wise rider, namely that householders in districts which had been designated as smoke-control areas should get financial compensation for the cost of converting their fires for smokeless fuel.

4. Before new industrial furnaces were built it would be necessary to have prior approval from the local authority, to ensure that the furnace would be able to comply with the Clean Air Act. This sort of measure was already embodied in several local acts and it was a forerunner of the present intention in Britain: that environmental impact assessments should be required before new factories are put up.
5. For the theme of this book the most interesting passages in the report concern the relationship between the Alkali Inspectorate and local authorities. We have already described how the alkali inspectors, largely on their own initiative, came to concern themselves with emissions of smoke, grit, and dust, which lay outside their statutory authority. Their advice was sought by local authorities. Policies for smoke abatement and for the abatement of noxious gases, after nearly a century of separate development, began to coalesce.

There was a danger in this coalescence. We have seen how parliament kept the alkali inspectors on a short lead, cautiously giving them authority to control only those emissions for which there was a prospect of effective remedies. There was no overlap between their statutory responsibilities and the responsibilities which local authorities could undertake to abate other kinds of air pollution. But the Public Health Act of 1926 entitled the inspectors to inspect works which were not scheduled under the Alkali Act. Under Damon's leadership they began to take an active interest in smoke abatement and because of their superior expertise they came to influence policies which were still the statutory responsibility of local authorities. At the same time some big local authorities were beginning to feel that they could serve their constituents better if they could take over from the alkali inspectors the control of those emissions which no longer presented difficult technical problems. In brief, there was a need to reconsider the boundaries of authority between the alkali inspectors and the local authority's sanitary inspectors.

The Beaver Committee tackled this question and proposed a clear and simple solution. It was that the Alkali Inspectorate should be responsible for controlling all emissions from industrial processes, including dark smoke, grit, and dust, where abatement presented special difficulties, and that the basis for control should be the use of the best practicable means.[17] The Committee suggested an initial list of processes which should be transferred to the alkali inspectors. They included metallurgical works, power-stations, gasworks, coke works, ceramic works, lime works. The list was to be reviewed annually, and as soon as the special difficulties were overcome, control of the processes should be 'made subject to the ordinary law', i.e. they should pass from the jurisdiction of the Alkali Act to the jurisdiction of the Clean Air Act, administered by local authorities. This solution implied an important change in the status of the alkali inspectors: they would concentrate upon unsolved problems of industrial air pollution; they would determine the criteria for best practicable means; and as soon as control of smoke, grit, and dust from a registered process came within the competence of the sanitary inspectors of a local authority, the process should be deregistered and pass out of the control of the alkali inspectors. 'The local authority inspectors and the central inspectorate should be regarded not as separate entities but as a partnership working together for the solution of a common problem...'[18]

One of the many merits of the Beaver Report is that it dotted 'i's and crossed 't's. It would have been unwise to propose smoke-control areas if there had not been a sufficient supply of smokeless fuel to be used in these areas. The Committee was 'entirely satisfied that...enough coke could be found to permit the creation of smokeless zones and smoke control areas as fast as the administrative action and replacement of appliances would permit during the next five years at least'.[19] But the Committee stressed that there must be a vigorous campaign of research and development to improve both the quantity and quality of smokeless fuels. It would have been unwise, also, to entrust local authorities with the administration of the Clean Air Act unless their inspectors had been adequately trained. The Committee was concerned that only 1.3 per cent of the total time of sanitary inspectors was spent on smoke abatement and most of them were not properly qualified to administer a Clean Air Act on the scale proposed in the report. Also it would have been unwise to create a Clean Air Act without providing for codes of practice, specifications for fuel appliances, and tests for smokeless fuels; responsibility for these was assigned to the British Standards Institution.

There were other examples of shrewdness in the report, notably an early attempt to calculate the economic cost of air pollution. The estimate—£250 millions a year—can be challenged on all sorts of grounds (see p. 144 below) but it did make a point that had never occurred to some people, namely that pollution really does cause serious material damage and massive economic costs.

The report was published in November 1954. There was no opportunity to debate it in parliament in January 1955; nor was there any immediate need to do so, for the Minister announced on 25 January: 'I can already inform the House that, while reserving their position in regard to individual proposals, Her Majesty's Government have decided in principle to adopt the policy recommended by the Beaver Committee.'[20]

The journey between a policy adopted in principle and its appearance on the statute book can be a long and weary one. In this case it was shortened by two circumstances. The first was a decision by the City of London not to wait for national legislation but to create its own smokeless zone by promoting a City of London (Various Powers) Bill. This was debated in parliament in March 1954, which gave MPs an opportunity to prod the government while the Beaver Committee was still deliberating.[21] One excuse still being used for taking one's time was an insufficient supply of coke. An MP for Manchester quoted a disclosure in the *Manchester Guardian* that there was an embarrassing surplus of a million and a half tons of coke left over after the winter 'with no prospect of getting rid of it unless the authorities made haste to establish many more smokeless zones'. He was followed by an MP for Salford (a satellite of Manchester), who quoted a letter from the Ministry of Housing and Local Government, written in 1952, turning down Salford's application for permission to create a smokeless zone on the ground that there was not enough coke, nor enough steel for the conversion of grates to burn coke; a challenge which drew from the Parliamentary Secretary for the Ministry (E. Marples) the assurance that 'now we have sufficient steel and smokeless fuel. It is a different kettle of fish.' This debate—which led to the approval of the Bill for the City of London to create a smokeless zone four months before the Beaver Committee reported—was a warning to the government that parliament had better not be kept waiting too long for a national Clean Air Bill.

Nabarro's Bill

The second circumstance which shortened the journey between agreement over the Beaver Report and a Clean Air Act was that Gerald Nabarro, a picturesque and flamboyant Tory MP, won the ballot which entitled members to present a private Bill to the House. He had decided to present a Clean Air Bill and the publication of the Beaver Report, only a few days before the first reading gave great impetus to it. He plunged into the campaign with characteristic panache. He is said to have had his outgoing mail stamped: 'Gerald Nabarro's Clean Air (Anti-Smog) Bill. 2nd reading, Friday 4 February 1955'.[22]

The debate on 4 February was a lively occasion. It went on from about 11 in the morning to 4 in the afternoon. Nabarro had been well briefed by the National Smoke Abatement Society for months beforehand. He began with some telling historical anecdotes, such as 'the earliest attempt to substitute smokeless fuel for raw bituminous coal', made in 1595 by Thomas Owen from South Wales, who had suggested that he should be allowed to bring anthracite coal to London—a suggestion which, if it had been adopted on a large enough scale, would have had a dramatic effect on London's smoke. Nabarro's Bill was, in his own words, 'Beaver, nearly the whole of Beaver, and practically nothing but Beaver'. For the drafting of the Bill he was able to follow the wording in the considerable number of local Acts already giving towns power to control smoke; indeed he made the good political point that what he was proposing might almost be considered a consolidating measure, putting into national legislation what already existed in much local legislation. The debate that followed was complimentary to Nabarro and, on the whole, favourable to the Bill. There were a few twitches of the sort of opposition that plagued poor Lord Stratheden and Campbell in the 1880s. Mr Higgs declared his instinctive reaction against a Bill which invaded peoples' houses: 'a Bill which imposes a penalty of that magnitude [he was referring to the proposed fine or imprisonment for a second offence] on a British householder for burning British coal in a British grate needs careful scrutiny'. 'I feel', he concluded, 'that the Measure is ill-timed and in the wrong form.'[23] Other MPs criticized the Bill as not being radical enough. It received influential backing from Enoch Powell, one of the more intellectually distinguished members of the House, who put it into its historical setting:

I believe that the method which the Bill adopts is in the line and tradition of our most successful public health legislation and of those successive great Measures which have dealt with the working and living conditions of our people over the last 100 or 120 years.

And Powell went on to define this tradition with Victorian sonority:

when a definite evil is ascertained, to the continuance of which public opinion is strongly adverse, then, as soon as avoidance and prevention become practicable, that evil is defined and a prohibition imposed upon it by legislation.[24]

When some major reform is contemplated in parliament, it often happens that it is introduced first as a private Bill; we have described many examples of this in the political history of clean air. If it is a reform which will involve public expenditure, there is no prospect that the private Bill will be allowed to pass into law, for the government has to be committed to financing the public costs of the reform, and will therefore want control over the wording of the Bill. So a common procedure is for debate on the private Bill to be encouraged (this gives the matter an airing and helps the government to find out how much support there is and what the difficulties are likely to be). The proponent of the Bill then announces at the end that he is willing to withdraw it, provided the government will introduce a Bill on similar lines; for then, provided the government has a majority, the measure is guaranteed to go through. That is what happened on this occasion. The Minister switched on a green light: there was, he said, no difference of opinion 'in this House and very little outside' about the intention of the proposals. But (he had of course been briefed by his civil servants to be cautious), he could not advise acceptance of the Bill in the form Nabarro had presented it, for it had been framed without consultation with local authorities which would have to administer it; its drafting was defective; and it contained no financial provisions necessary to make it effective. 'Honourable Members have to choose', he said, 'between a small bird in the hand and a bigger bird in the bush.'[25] So, with the promise that a government Bill would be put forward on the lines the Beaver Committee recommended, Nabarro withdrew his Bill.

The Clean Air Bill

The government kept its promise. A Clean Air Bill was given a first reading (a formality with no debate) five months later, on 26 July. It came up for a full debate at the second reading on 3 November.

The Bill, as presented by the government, differed from Nabarro's Bill in a predictable way. If Nabarro's Bill was 'practically nothing but Beaver', the government Bill was Nabarro's Bill with all the abrasive edges filed down and burnished to make it acceptable to the three interests without whose co-operation it stood no chance of success, even if it were put on the statute book. These interests were the civil service, industry, and local authorities. Over a reform as far reaching as this one, the civil servant is apt to devise problems to put in the way of every solution proposed by reformers: he has to make the new law work, and he will therefore be at great pains to rid it of any difficulties in administration and enforcement. The industrialist has two motives in trying to water down a clean air act: he wants to minimize its cost to himself in trouble and cash; and he is aware, as the parliamentary draftsman cannot be, of the complexities (such as the inevitable emission of smoke when a boiler is started up) which would make a law unworkable unless they were foreseen. Local authorities sensitive to the sovereignty of the ratepayer and aware that they may not have the expertise to deal with some of the technical problems of air pollution, are apprehensive lest their powers should be weakened by national legislation which would deprive them of their influence over local affairs. So it was no surprise that the government Bill was more circumspect than Nabarro's private Bill. Nevertheless it did adopt the important innovations in the Beaver Report. It prohibited the emission of dark smoke, defining 'dark' by a shade on the Ringelmann chart, as the Report had proposed. It required that new furnaces should be designed to be 'so far as practicable smokeless'. It required furnaces above a certain size to be fitted with equipment to extract grit and dust. It empowered local authorities—subject to confirmation by the Minister—to declare part or whole of their districts to be smoke-control areas, where the only fuels permitted would have to be smokeless fuels; and it provided for generous grants to cover the cost of converting domestic grates to burn smokeless fuels. It adopted the Beaver Committee's recommendation that the Alkali Inspectorate should be given control over dark smoke, grit, and dust from industrial premises where abatement still presented technical difficulties; and it provided for the transfer of this responsibility to local authorities when the difficulties had been removed. The old requirement, to prove that smoke was a nuisance (under the Public Health Act, 1936) before it could be suppressed, had almost disappeared. But not quite: smoke not dark and not emitted from a private dwelling had to be deemed to be a nuisance before anything could be done to abate it, and the offender was protected if he could establish that he had used the best practicable means to prevent the nuisance.

Such was the gist of the Bill debated in the House of Commons on 3 November 1955. The debate went on for six hours, though it has to be added (lest the length of the debate be taken to imply intense parliamentary interest) that most of the speakers came from constituencies where clean air was a political issue. There was no risk that the Bill would not pass, though there were several amendments before it was put on the statute book. The only criticism of the Bill from a supporter of the government came from Nabarro.[26] For him the Bill was not tough enough: it was, he said, ill-drafted; too leisurely (it gave industrialists seven years' grace before the dark smoke clause began to bite, Nabarro wanted only three years); and with wide escape routes. The hand of the Federation of British Industry, he said, 'is writ large between the lines'. No doubt he was right: he was a member of the FBI himself.

As for members of the Labour Opposition, it was of course their business to pick holes in the Bill. We can disregard the rhetoric of denigration as mere political ritual: it was called 'a miserable hypocritical shadow of a Bill', 'a lifeless, spineless, Measure', 'a little mouse of a Bill'. But some of the criticisms were very much to the point; it was indisputable that a lot of concessions had been made in order to make the Bill palatable. Thus, an industrialist could get away with emissions of dark smoke not only when lighting up his furnace, but also if there had been an unforeseen failure of his equipment, or if he had used unsuitable fuel at a time when suitable fuel was unobtainable and had taken 'all practicable steps' to prevent the dark smoke. He could get a certificate of exemption during the seven years of grace if his emissions of dark smoke were 'due to the nature of the building or its equipment' or because 'it had not been practicable to alter or equip the building'. Local authorities, too, had been placated; they were under no obligation to create smoke-control areas; so there was no guarantee that domestic smoke would be abated at all by the proposed law. The alkali inspectors came out of the debate very well supported. The Beaver Committee had applauded their work,[27] and the Minister (Duncan Sandys) when he came to close the debate, made the point that very few local authorities would be equipped to take responsibility for the kinds of air pollution for which there were no reliable controls and which had to be managed by the best practicable means. These had to be left to the expertise of the Alkali Inspectorate: 'These inspectors are', he said, 'very rare birds.'[28]

After this long debate the Bill went to a committee which met thirteen times and grappled with the detailed wording. It was another opportunity for the three interests—civil service, industry, and local authorities—to make the Bill as comfortable as possible for themselves. Local authorities, for instance, were concerned to protect their interests against those of the alkali inspectors. Should the clause providing for transfer of processes to the local authorities read 'The Minister *shall*...' or 'The Minister *may*...'? Should local authorities, if they did take over control of a process formerly scheduled under the Alkali Act, have the powers of alkali inspectors? And there was a great deal of to-do about practicable means. Should the formula be 'best practicable means' or 'any practicable means'? And should a definition of the formula take into account the cost and financial consequences of abating air pollution? Or should a measure so important for the nation's health be carried through irrespective of cost? Also, should the criterion of what is practicable take account of local conditions (an echo of the famous pronouncement a century ago by Mr Justice Thesiger (which we quote on p. 51). Argument about this was still going on when the Bill, having survived the committee stage and a third reading, went to the House of Lords. There the Lord Chancellor (L. Kilmuir) drew attention to the ancestry of the clause which used the word 'practicable'. 'The essence of the definition', he said, 'is that the word is to be construed as meaning not only what is physically practicable but what is reasonably practicable; and that, it is thought, must necessarily entail having regard to local conditions and circumstances and the financial implications amongst other things.'[29] Despite attempts to narrow it, the definition of 'practicable' in the Bill remained broad in the final Act, to mean:

> reasonably practicable having regard, amongst other things, to local conditions and circumstances, to the financial implications and to the current state of technical knowledge, and 'practicable means' includes the provision and maintenance of plant and the proper use thereof.[30]

The Minister had promised a 'bigger bird in the bush' to replace Nabarro's 'small bird in the hand'. At the time the Minister's bird seemed smaller than Nabarro's, but this may well have improved its capacity to survive. Ever since the introduction of the first Alkali Act in 1863 politicians had realized that penal clauses in legislation to protect the environment are useful only as a last resort; the only hope of success for such legislation is willing consent from the people who are expected to obey it. This co-operation has to be purchased by an attitude of reasonableness on the part of the legislators. During the debate on the Bill, the Lord Chancellor reminded the House: 'There is always a real danger, in trying to cure ills by creating offences, that what is caused is a disrespect for the law'.[31] It was the need to frame a law which would be obeyed by consent rather than by compulsion which brought upon it such epithets as 'hypocritical', 'spineless', and like 'a little mouse'.

The response of local authorities

The Clean Air Bill received the royal assent on 5 July 1956. Today, twenty-five years later, the levels of smoke, grit, and dust in the atmosphere are only a fraction of what they were in 1956. How much credit does the Clean Air Act deserve for this dramatic improvement? Not all the credit. The severity of air pollution had been declining in Britain for some years before 1956; this was due to the efforts of the alkali inspectors, and to the initiative of local authorities which had already brought in their own measures to control smoke; it was due also to the change-over to gas and electricity, the spread of central heating, shifts in population from city to suburb, and other circumstances. However, even when all these circumstances are taken into account, the verdict voiced by some critics—that the Clean Air Act was a failure because it was too permissive—is unjust. It went as far as consensus would follow. It was the section about smoke-control areas which aroused expectations from the public. Let us examine what happened about these.[32]

The Beaver Committee, aware of the gigantic administrative task of controlling smoke on a national scale, recommended that a start should be made on the 'black areas' of Britain, which it illustrated by a map attached to its interim report; but it did not specify which local authorities should be classified as being in black areas, nor were black areas mentioned in the Clean Air Act. It would have been economically impossible for every local authority, simultaneously, to be permitted to create smoke-control areas—for one thing, there was not enough smokeless fuel to supply all areas—so the initiative lay with the Ministry of Housing and Local Government to advise local authorities whether they were 'black' or not. This it did not do until January 1959, thirty months after the Act was put on the statute book. The circular issued in January 1959 requested 325 local authorities to prepare a phased programme for establishing smoke-control areas in the following five years.

The replies were not encouraging. In 1962 a second request was issued to local authorities. The smoke-abatement lobby was getting restless; were the sceptics of 1956 going to be proved correct after all? A few months later came a further setback. On 24 May 1963 in reply to a parliamentary question, the Minister for Housing and Local Government confessed:

whereas hitherto it had been possible to rely on open-fire gas coke for domestic grates...in smoke control areas, a major increase in the production of gas coke could no longer be expected owing to the rapid technological changes now taking place in the gas industry.

At once the applications from local authorities dried up. To create smoke-control areas was not everywhere a popular move, especially in some industrial areas where smoke was still equated to employment; so an excuse for deferring a decision was welcomed by some councils. By 1965 the prospect for the supply of smokeless fuels had improved. The Ministry issued a revised report and this was followed by a third request early in 1966 to those selected local authorities in black areas which had not responded to the Ministry's first two circulars. This time the circular rattled the bones a little: in the light of replies to his circular, the Minister would be able to decide whether or not 'to make smoke control a statutory order'. This was the whispered *dirigisme* of a Labour government, which had by now taken the place of the Conservative one.

Hard on this appeal to local authorities to act came another discouragement: a curb on public expenditure which again dissuaded local authorities from creating more smoke-control areas. Some notoriously black areas, especially those where miners refused to forego their traditional free allowance of coal, continued to resist the government's persistent requests. At last it seemed desirable to embody the gentle *dirigisme* in statute. This was done not by government Bill but by a Private Members' Bill moved by Mr Ian Maxwell, who had some highly polluting brickworks in his constituency.[33] It was debated in February 1968.[34] 'Not a harassing Bill', Mr Maxwell said; its objects were to empower the Minister to prescribe emission limits for grit, dust, and fumes; to oblige a wider range of furnaces to install arrestors; to require local authority approval for the heights of chimneys; and—the really important clauses—to give the Minister power to direct local authorities to submit smoke-control programmes and to 'require them to carry them out'. The Bill was welcomed with practically no criticism. Indeed at the beginning of the Committee stage nothing was said about the content of the Bill; instead, eight columns of Hansard were squandered on the record of a wrangle about the propriety of giving a Private Members' Bill such a rare precedence! Despite some hold-ups contrived by the Opposition, the Bill became law (as the Clean Air Act 1968) on 25 October:[35] a reasonably swift passage for a statute which, in other circumstances, would have been highly controversial.

But scarcely was this new law on the statute book than a fresh famine in smokeless fuel put a stop to its use; indeed the government had to suspend programmes for smoke control in regions where they were already under way. The permissiveness of the Clean Air Act of 1956 was not the cause of this halting start; the fault was the lack of any parallel legislation over the years 1956-68 to create, and if necessary subsidize until demand rose, an industry to supply the necessary quantities of smokeless fuel. It was the

Ministry of Fuel and Power which had blundered. Supplies of gas coke fell from 22.7 million tons in 1958 to 3.7 million tons in 1970. In the early 1970s some gas plants were kept working not primarily to supply gas but to supply the by-product, smokeless fuel.

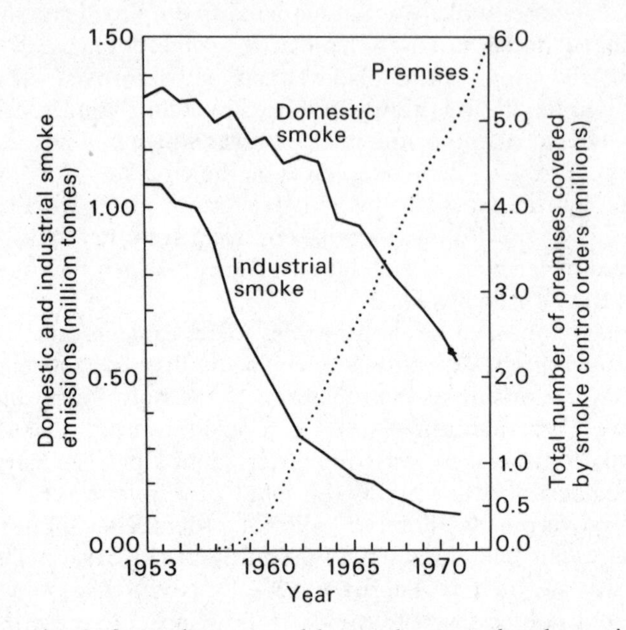

Fig. 3. Comparison of premises covered by smoke control orders, with domestic and industrial smoke emissions, 1953–1973. Department of the Environment, *Clean Air Today*. HMSO, London (1974).

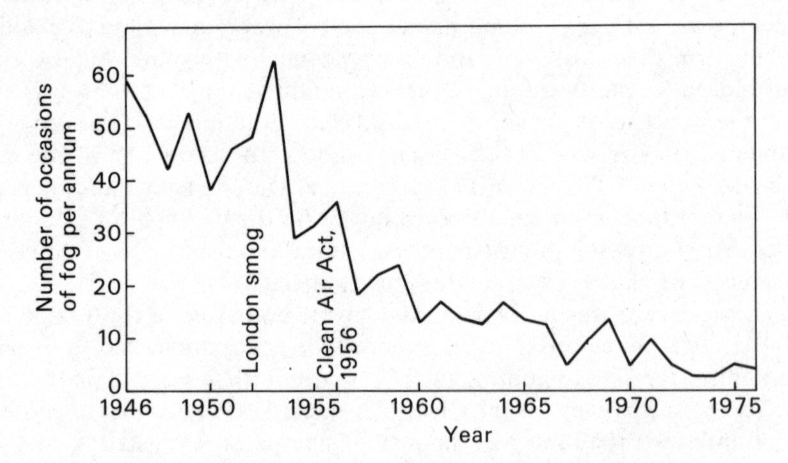

Fig. 4. Number of occasions of fog (visibility less than 1 km) at 09.00 GMT at the London Weather Centre, 1946–1976. Data by courtesy of the Meteorological Office.

Despite all these shufflings and setbacks, the record of smoke control, by the time the Royal Commission on Environmental Pollution came to look at it in 1971, was impressive.[36] The achievement is summarized in Figs. 3 and 4. Over 5 million premises were covered by smoke control orders. Domestic smoke had fallen from emissions of about 1.35 million tonnes a year in 1956 to about 0.55 million tonnes, and industrial smoke from about a million tonnes to about 0.1 million. Even though the Clean Air Act itself cannot take the credit for all this, the credit is certainly due to the evolving social values which were crystallized out, so to speak, in the Clean Air Act. As we have had occasion to note before, the part legislation played was to make explicit what was already implicit in the minds of people who influence public opinion. It was often the health inspectors themselves, local government officials, who nagged away at their political masters to put in applications for smoke-control orders. Naturally, since the decision was left to local authorities (even after the 1968 Act the Minister did not make much use of his power to intervene), the health inspector could not persuade his councillors to put clean air at the top of the priority list if there were more urgent schemes, such as slum clearance, to be financed first. This explains the uneven, and perhaps inequitable, distribution of pollution control over the country.

And what, over this same period of the 1950s and 1960s, was the state of progress under the alkali inspectors?

10

A UNIFIED AIR POLLUTION INSPECTORATE

The Alkali Inspectorate and the Clean Air Act

At the end of Chapter 8 we brought the story of the Alkali Inspectorate up to the 1950s. We then turned to consider progress on smoke abatement which began after the London smog of 1952. We now take up again the story of the Alkali Inspectorate.

The genial and effective W.A. Damon retired from the office of Chief Inspector in 1955, though he was retained as consultant for another four years. He visited New Zealand to advise the government there and he helped his successor in special assignments, some of which arose as a consequence of the Clean Air Act.

Damon's successor was J.S. Carter. His tenure, from 1955 to 1964, was a phase of consolidation and adaptation of the Inspectorate to new developments in industry. The most significant political events which affected his work were the passing of the Clean Air Act in 1956 and the issue in 1958 of an Order which greatly extended the responsibilities of his small team.

The Clean Air Act, as we described in the last chapter, ended the old dichotomy between smoke abatement and the abatement of noxious gases, and put in its place a new and more logical dichotomy. Industrial emissions which could be reliably abated by known methods were normally to be controlled by local authorities; industrial emissions for which there was no simple and tried mode of abatement and which still presented technical problems were to be registered under the Alkali Act and controlled by the alkali inspectors. And, as the Beaver Report had emphasized, there was to be a working partnership between the local authorities and the Inspectorate. This brought smoke, grit, and dust emissions from power-stations and the like—about which the Alkali Inspectorate had given extrastatutory advice since Damon took an interest in the problem in the 1930s—into the direct responsibility of the Inspectorate; it was something Angus Smith had foreseen as far back as 1870. The Act provided also for a transfer of responsibility in the other direction, from the Inspectorate to local authorities. Effect was given to this in 1960, when the corporations of Liverpool and Manchester, and the Clean Air Committee of Sheffield and District, were permitted to make themselves responsible for controls handed over to them by the Alkali Inspectorate.[1]

The procedure for adding noxious gases or processes to the schedule under the Alkali Act had been simplified by a section in the Public Health

Act, 1926, which we mentioned on p. 95. Under this section it was necessary only for the Minister to make an Order which had to be laid before parliament but which did not have to go through the formalities of legislation. Nine such Orders were issued between 1928 and 1971. Some were to add new chemical industries to the schedule (e.g. works making acrylates, which are highly pungent); another (issued in 1966) made minor amendments and consolidated seven earlier Orders; other Orders (e.g. one issued in 1971)[2] de-scheduled some processes which no longer needed the expertise of the inspectors, and could be returned to the control of the local authority. The technical details of these Orders are of concern to the chemical engineer; it is their political overtones which are our concern.

At the ground level, so to speak, there was, on the whole, cordial co-operation between alkali inspectors who were servants of the central government and public health inspectors who were servants of local authorities. But in the political stratosphere, far above the day-to-day problems of smoke abatement, all was not sweetness and light between the inspectors' masters. Evidence of this came to the surface in the public inquiry which had to be held before the first Order was issued after the 1956 Clean Air Act. The enlarged responsibilities of the Alkali Inspectorate—to take under its wing industrial emissions of smoke, grit, and dust if these were difficult to control—required considerable additions to the schedule of registered works. The public inquiry opened in May 1957. It was conducted by Sir Frederick Armer. Hearings went on for nine days. Evidence was taken from the Iron and Steel Federation, the Steel Founders' Association, the National Association of Non-Ferrous Scrap Metal Merchants, Imperial Chemical Industries, the Gas Council, the National Federation of Clay Industries, and other similar polluting industries. Without exception they wanted to have some or all of their processes scheduled under the Alkali Act, subject to control by the alkali inspectors. Their motive was not altruistic: it was that they knew they could not comply with the provisions of the Clean Air Act. Armer had to satisfy himself that there were special technical difficulties in the prevention of pollution before he would recommend that any of the processes should be so scheduled; he rejected some applications (e.g. for one treatment of iron castings, he wrote: 'It can hardly be said that the solution of this smoke problem presents any special technical difficulty... I do not consider that a case has been made out for bringing this process under the control of the Alkali Act'). The only processes fit to be scheduled were, in Armer's view, ones which local authorities would not be competent to control. His criteria, as he repeatedly said in his report, were purely technological. But representatives of some local authorities, who also took part in the hearings, added a political dimension to this inquiry. The Association of Municipal Corporations, while conceding that 'there are some industries which present technical difficulties upon which the advice of experts is necessary', felt that so far as possible control should be vested in local authorities. The Association then disclosed its real concern, which was more about politics than about the state of the environment: if the proposals from the industrialists were all embodied unmodified in an Order, it would, the Association said, 'have the result of an unwarranted diminution in the powers of local authorities'.

The Order issued in 1958 increased the number of registered works from 872 to 2160 and the number of separate processes from 1733 to 3412.[3] The statutory responsibilities of the Alkali Inspectorate were greatly increased, and its strength more than doubled. The number of districts in England and Wales was increased from seven to twelve;[4] and by 1960 the Chief Inspector was assisted by 2 deputy chief inspectors, 12 district inspectors, and 10 inspectors: an increase of 16 persons.[5] In Scotland an inspector was appointed to assist the Chief Inspector there.[6] But a secondary and unwelcome consequence of the greater responsibilities was that the Alkali Inspectorate found itself exposed to the unwelcome attentions of politicians in the old rivalry of interests between central and local government. The Inspectorate (like the universities) preferred to keep out of the news and to get on with its work undisturbed by questions in parliament. In the 1960s this comfortable insulation was withdrawn (as it was for the universities, too).

The first sign was a sprinkle of parliamentary questions. Was the Alkali Inspectorate equipped to undertake the new duties imposed on it by the Clean Air Act? How many inspectors were there? How much were they paid? When the Order containing the recommendations Armer had made was laid before the House of Commons, Nabarro moved that the Order be annulled.[7] Of course he did not really want to annul the Order: it was one of the first fruits of the Clean Air Act he had done so much to promote. But he wanted to make sure that the Act was interpreted according to sound Conservative doctrine: namely that the maximum responsibility should be delegated to local authorities. Nabarro criticized the Order as 'ham-fisted', taking away 60 per cent of the enforcement powers of the local authorities (which was untrue),*making the Alkali Inspectorate the senior partner and the local authority the junior partner (which was a distortion of the truth). He went on to assert that the Inspectorate was a weak body: even when enlarged to a strength of 28, as proposed, the combined salaries of the inspectors would be no more than £47 730 p.a. The Order was a foray in empire-building by central government 'which ought to be resisted by all who have the interests of strong local government at heart'. Nabarro got some support for his views; the Minister stoutly defended the Order and corrected some of the misstatements that had been made. Nabarro was content: he withdrew the motion for annulment. The issue had been aired. The Conservative philosophy had been re-asserted.

*Over 25 000 works emitting smoke were still left under the surveillance of the local authorities.

In 1962 another Order was laid before parliament. Again, in order to give parliament a chance to comment on progress in the campaign for clean air, a member (Barnett Stross, who represented a division of Stoke on Trent, a city with a notorious record for polluted air) moved that the Order be annulled.[8] This time the intention was more welcome: it was to put on record appreciation for the work of the alkali inspectors in Stoke and to encourage the Minister for Housing and Local Government to support the Inspectorate. There was no criticism to answer this time, possibly because the processes to be brought under control included uranium works; no local authority wanted to be saddled with control of radioactive materials. The public inquiry lasted only 16 minutes.

The Alkali Inspectorate reacted sensibly to this growing interest in its affairs. Faithfully every year, Carter reported the number of complaints his inspectors had received. He never ducked the complaints. Over steelworks, added to his schedule in 1958, he confessed there were major problems to be solved before the nuisances could be abated. Power-stations, he recorded, 'will present a major problem for many years to come'. More complaints were received about ceramic works than about any other class of manufacture. As to nuisances from lime works, the outlook was 'gloomy [but] not entirely hopeless'.[9] He noted that the complaints had to be regarded as a 'measure of public awareness for air contamination'. In the past, for instance, heavy pollution from coke oven plants had been accepted as a matter of course; now 'there are signs that this attitude of mind is disappearing'.[10] Complaints about the smoke and smell from paraffin oil works were increasing, but—Carter records—this was because more people were becoming aware of the Clean Air Act and of the steady increase in smoke control areas. He had the familiar experience that improvement in an amenity is a mixed blessing: it is likely to increase expectation more than it increases satisfaction. The designation of an area to be a 'smokeless zone' he recorded, 'almost invariably brings increase in local complaints'.[11] And over iron and steel works, 'as smoke control areas grow, public opinion is going to become more and more impatient and insistent' that these works should cut down their pollution.[12] The Inspectorate had not the time, or the money, to indulge in exercises in public relations. Carter did, however, try to respond to the growing public interest in his reports (though they still got very little notice in the press), by including, from 1960, historical reviews of specific pollutants. These give a vivid impression of the range and complexity of the problems the inspectors had to tackle. The inspectors recorded also their lighter moments: one was a request for advice about a yellow deposit on the ground over a wide area adjacent to an oil refinery. It turned out to be pollen!

In 1963 the Alkali Act was 100 years old. The original sole duty of the inspectors, to reduce by 95 per cent emissions of muriatic acid from works using the Leblanc process, still remained one of their duties; though the process was on the way out and the last plant closed in 1964: a quiet death for a manufacture which had once dominated the chemical industry in Britain. To this basic duty an extraordinary variety of new tasks had been added, tasks undreamed of by the early inspectors: dealing with man-made fibres, bromine and magnesium from sea water, plastics, nuclear fission. There had been successes such as the dramatic reduction of smoke from potteries and—at great expense—of smoke and grit from power-stations. There remained unremedied nuisances: dust from cement production, fluorides from brickworks and dirt from carbon black works. The costs of abating pollution were already considerable: for a 2000 megawatt power-station the capital cost of equipment alone (a 650 foot chimney, dust extraction plant, and dust collection plant) amounted to £3.5 millions. There was increasing and cordial liaison between the inspectors and local authorities (though, as a royal commission was to discover ten years later, the partnership hoped for by Beaver was not everywhere successful). Left alone to get on with a job few of the public had ever heard of, the alkali inspectors had earned the respect of industry and had secured widespread compliance with their standards by persuasion rather than injunction. Looking at the historical scene at its centenary, Carter wrote:

> At first sight there seems no basis for comparison between 1863 and 1963. Yet the fundamental requirement that industry shall use the best practicable means to render its emissions harmless and inoffensive is timeless and is independent of the nature of the process whether chemical, metallurgical, combustion or nuclear. Proof of this timelessness is supplied by the fact that the last five years [1959-63] have seen a greater expansion of the scope of the Act than had the previous 95.[13]

The Inspectorate on the defensive

In Carter's time as Chief Inspector, one heard only the overture of mass concern for the environment. His successor, F.E. Ireland, was exposed to the full chorus, amplified by the media and exploited by journalists into a sensational threat to industrial society. *Silent spring*, published in 1962, was an eloquent indictment of the over-use of pesticides. A depleted oxygen level in Lake Erie, polluted by wastes from detergents and agriculture, provoked the hysterical announcement that 'Lake Erie is dead'. Mercury poisoning from fish in Japan, damage to forests from acid rain in Sweden, oil spills at sea, toxic wastes dumped in river beds, radioactive contamination: as soon as one alarm was out of the headlines another took its place. It is impossible to say, in the clamour which culminated in the UN Environment Conference in Stockholm in 1972, just how widespread the mass concern was. In any case it has not turned out to be durable; by the mid-1970s much of it had evaporated, to be replaced by concern about energy supplies. But the clamour had some durable consequences. One was the growth in influence of environmental lobbies and public-interest pressure groups. These are the successors to the lobbies that played such an important part in the smoke-abatement campaigns we have described. This culminated in a take-over of the word 'ecology' (coined in 1873 to distinguish a new discipline in biology) as a label for a new political party. Another consequence was the demand made by these lobbies for the public's right to know the facts about pollution and to participate in decision-making about the environment. 'Accountability' (a word appropriately put between quotation marks, for many who use it cannot define it) has become a postulate for our style of democratic government. We are now, in the 1980s, trying to work out patterns of accountability and participation to meet this demand.

The alkali inspectors could not have expected to escape from this current of change, though there were signs of pained surprise in some of their reactions to it. It was a very trying term of office for the Chief Inspector, for the alkali inspectors not only came under a good deal of criticism—some of it irresponsible and unjust—but they had to try to maintain their exacting presumptive standards of best practicable means in conditions of almost unprecedented economic stress. Between 1967 and 1975 the responsible ministers were pestered with questions in parliament. Why were so few industrial polluters prosecuted? Why, on the rare occasions when there were convictions, were the fines so derisory? Why were the inspectors so secretive about the data they collected and the standards they prescribed? What instructions were given to the Inspectorate about the interpretation of best practicable means?

The Clean Air Council—an advisory body created under the Clean Air Act under the chairmanship of the Minister—did its best to mediate between the alkali inspectors and their critics. A journalist, Jeremey Bugler, wrote a paperback containing waspish attacks upon the Inspectorate as a body of secretive officials, not accountable to the public nor even to their own Minister, hand-in-glove with the industrialists whose pollution they were paid to abate. In June 1972 the Clean Air Council invited Mr Bugler to defend his criticisms.[14] It was put to him by members of the Council that his presentation of the case was unbalanced, exaggerated, and a travesty of the facts. Apparently Mr Bugler did not deny this. The book, he said, was not intended to be objective or balanced: it was 'campaigning journalism'! It is no wonder that some passages in the Chief Inspector's annual reports seem in hindsight (always a dangerous perspective) to be a trifle querulous and defensive.

Official reports are, of course, no match for strident pamphleteering and it was not long before a member of parliament, more familiar with the pamphlets than with the reports, brought in a Bill to reform the Alkali Inspectorate. He was Mr N. McBride, an MP for Swansea. The intention of the Bill was to oblige the alkali inspectors to become more militant. He spoke of the 'mounting criticism' of the Inspectorate, the 'cloak of anonymity' it had worn since the passing of the Alkali Act in 1863, the need for the Inspectorate to 'emerge from the Victorian shadows where it has been lurking willingly for far too long'.[15] (Mr McBride omitted to mention that the anonymity had been breached and the shadows illuminated by some eight thousand pages of published annual reports since 1863!) He reported the criticism that the Inspectorate 'is in complete ignorance of the cost of air pollution control'.[16] (He must have overlooked the annual report for 1968, which carries a table setting out the costs of air pollution control for ten industries—an overall total for the ten-year period of some £480 millions—broken down into capital costs, working costs, and research and development.)[17] Mr McBride was supported by a fellow MP for Swansea, Mr Alan Williams, who spoke movingly about the 'sheer weakness and ineptitude of the inspectorate'.

The Bill called for compulsory consultation with local authorities before standards were set for emissions; it wanted the Inspectorate to be given new powers to require all industrial plant to conform to specified (not just presumptive) standards for emissions of wastes or toxic gases; it proposed that industrialists should be obliged to deposit 'performance bonds' which they would forfeit if pollution from their works did not comply with the specified fixed limits; it proposed also that data about emissions from individual works should be published (this last proposal was something the Aberdare Commission had recommended a century before).

A few clauses in the Bill, if they had been adopted,would have improved the efficacy of the Inspectorate; but the Bill as a whole, if it had been passed, would have been a disaster. It showed a total disregard for the historical roots of the service. It would have turned the inspectors into adversaries instead of advisers to industry. The Minister courteously corrected some of the more irresponsible statements made in the debate, defended the use of best practicable means as a better formula for control than statutory standards would be, and closed by saying that 'to make a change in the basic principle on which the Inspectorate works would be a mistake'. But he did say—and he was right—that there could be an improvement in communication; this is a point we pick up when we come to assess the work of the Inspectorate.

Take-over by the Health and Safety Executive

The Bill was dropped but the criticism went on. Some of it, as we discuss in the next chapter, was valid criticism; but it was ineptly made and this gave the Inspectorate an excuse for disregarding it. Meanwhile a fresh idea came to the surface in a parliamentary question: the suggestion that the Alkali

Inspectorate should be merged with the Factory Inspectorate. A much publicized incident gave support to this idea. In 1968 a very large smelting works for making zinc by a new process was opened at Avonmouth, near Bristol. In 1971 there was evidence of lead and cadmium escaping from the works into the surrounding countryside. This raised problems of divided responsibility. The factory inspectors were responsible for protecting workers inside the premises from lead and cadmium pollution. If the lead and cadmium escaped from the premises, their control became the responsibility of the Alkali inspectors. While discussions about the Avonmouth works were going on a committee under Lord Robens was examining the legislation for safety and health at work. The Committee published an important report in July 1972, upon which the present law is based.[18] Two features of the report are of critical importance for our story. The first was an examination of the pressures then being brought to bear on inspectors to adopt a more militant attitude toward industrialists who failed to comply with standards, by setting fixed standards and using the courts to enforce them. The report dismissed this attitude as misconceived, and came down firmly on the side of those who used the flexibility of best practicable means and relied on persuasion for getting co-operation from industry.[19] It was a complete vindication of the style of working of the Alkali Inspectorate.

The second critically important feature of the report was a recommendation that a new authority responsible for safety and health at work should be created, and that the Alkali Inspectorate should be moved from the Department of the Environment (the super-department which had subsumed the Ministry of Housing and Local Government) to the Department of Employment.[20] The government accepted this recommendation, and as soon as the legislation based on the Robens Report went through parliament, to emerge as the Health and Safety at Work Act, 1974, the transfer was made. The Inspectorate (whose formal title had been amended in 1971 to H.M. Alkali and Clean Air Inspectorate) was lifted—to its undisguised dismay—from the Department of the Environment and put under the Health and Safety Executive of the Department of Employment.

The transfer took place on 1 January 1975, in the face of a forthright recommendation to the contrary from the Royal Commission on Environmental Pollution. In a report on air pollution (discussed below, p. 129) the commissioners wrote:

> Our firm conclusion is that the incorporation of the Alkali Inspectorate in the Health and Safety Executive organisation is potentially damaging to the interests of the environment... the Alkali Inspectorate should be removed from the Health and Safety Executive forthwith and should return to the direct control of the Department of the Environment.[21]

It is too early to say whether this change will turn out to be as misconceived as many people think it is. There is no doubt whatever that in one respect it was a misfortune. The Alkali Inspectorate, set up to protect people, property, crops, and domestic animals from the hazards of polluted air, was acquiring a new and important purpose: to protect the air from the hazards of over-exploitation by people—a rudimentary empathy for the environment itself. It was therefore appropriate that the Inspectorate should be in a department charged with concern for the environment, and which has administrative responsibilities for local authorities which deal with solid waste and for regional water authorities which deal with sewage.

This was the emphatic view of the Standing Royal Commission on Environmental Pollution, set up in 1970. In its first report, a survey of the state of the environment in Britain, the Commission saw no need to change the status of the Alkali Inspectorate. Indeed, the Control of Pollution Act, 1974, based on the Commission's findings, had scarcely anything to say about air pollution because the needs for legislation were already well covered by the Clean Air Act and the Alkali Act. However, before the Health and Safety Executive took over the functions of the Alkali Act, the Inspectorate was submitted to one more intensive scrutiny.

It arose in this way: in the spring of 1974 a public-interest group calling itself Social Audit published a report on the Alkali Inspectorate which criticized severely the style of working of the Inspectorate.[22] On 23 May 1974 the Secretary of State for the Environment was asked whether, in view of the report by Social Audit, he would increase the number of inspectors and ensure that more information about their work was given to the public. The Secretary of State replied that he intended to ask the Royal Commission on Environmental Pollution to set up an urgent review of the whole system of air pollution control. The terms of reference to the Commission were:

> To review the efficacy of the methods of control of air pollution from domestic and industrial sources, to consider the relationship between the relevant authorities and to make recommendations.

The purpose of the inquiry was to dispel the odour of criticism which had for five years or more surrounded the bodies responsible for abating pollution of the air. It was in any case a timely decision. There had been no major inquiry into the matter for twenty years. During these twenty years an immense amount of information had been collected, notably in a national survey of air pollution published in three volumes in 1972, and in a sophisticated cost-benefit analysis of the damage caused by air pollution, published by the Programmes Analysis Unit in 1972.[23] There had also been major political and administrative changes affecting clean air policy: the creation of a Department of the Environment, the reorganization of local government, and Britain's entry into the European Community.

Her Majesty's Pollution Inspectorate

The Royal Commission reported in January 1976.[24] Its conclusions are the climax of our story, for they constitute a verdict on the policies which had evolved since Michael Angelo Taylor's committee reported on smoke in 1820 and Lord Derby's committee reported on noxious vapours in 1862. The Commission made a crystal-clear recommendation that the two independent policies of control—the one for smoke and the other for noxious gases—should now be combined into a single policy under a national pollution inspectorate. More than that: the Inspectorate, to be known as Her Majesty's Pollution Inspectorate (HMPI) should be created 'to ensure an integrated approach to difficult industrial pollution problems at source, whether these affect air, water or land'. HMPI would be centrally administered; it would comprise a small, highly qualified body 'which would focus on any industrial processes and plants creating difficult pollution problems'. Its expertise would be available to regional water authorities, which would remain responsible for pollution control in rivers, and to local authorities, which would remain responsible for the disposal of wastes on land. This new kind of Inspectorate would, if it were created, overcome the chief weakness of a piecemeal policy where there is no formal liaison between the different bodies responsible for abating pollution in the air, controlling effluents into water, and disposing of wastes on land. It could also influence manufacturers not just to get rid of their wastes in the way that gives them least trouble, but to try to prevent the wastes being released into the environment at all.

We said that the Royal Commission's conclusions are the climax of our story. This is because the Commission says what is really the last word that needs to be said about the formula which has woven its way through the history of policies for clean air: the 'best practicable means'. It is no secret that some members of the Commission were very sceptical about the value of this formula when they began to take evidence. As well they might be, for it had been pilloried in the press by people who did not understand how it was used, and little had been said, outside the Inspectorate's official publications, about its virtues. But as they informed themselves about the way the formula was applied, the commissioners changed their minds.

They were—and rightly—unhappy about some shortcomings in the way best practicable means had been used as a criterion. But they were unambiguous in saying that the criterion is 'inherently superior to control by nationally-fixed and rigid emission standards'.[25] They recognized the formula for what it is: a subjective (and to this extent vulnerable) cost-benefit balancing act which takes account of other things beside the simple abatement of pollution. Some members of parliament would have liked to see presumptive standards for emissions laid down in instructions from the Minister. This would look tidy but it would in fact waste money and resources, for a blanket-standard would be too severe for some works, and therefore unnecessarily costly, and too lax for other works, and therefore inadequately abating the level of pollution. So the Commission left the responsibility where it had rested for over a hundred years: the Chief Alkali Inspector, on the advice of his colleagues, decides what are the best practicable means in all the circumstances. His discretion is limited only by the handful of emission standards set by parliament in the nineteenth century.

The Royal Commission accepted this broad interpretation of best practicable means but it went on to spell out its implications. The inspectors were having to make decisions about things which lay outside their professional expertise. They were not accountants, able to judge whether a works would go out of business if they insisted on a severe presumptive standard. They were not biologists, able to assess a claim that some waste gas was injuring crops. The inspectors needed to enlist much more help than they had in the past from experts in disciplines other than their own.

Notwithstanding these cautions, the final verdict of the commissioners is evident in their recommendation for the unified pollution inspectorate, HMPI, they wrote:

> would seek the optimum environmental improvement within the concept of "best practicable means", employing the knowledge of industrial processes and many of the present techniques of the Alkali Inspectorate to reduce or modify the wastes produced, whether solid, liquid or gaseous. In effect, we have in mind an expansion of the concept of "best practicable means" into an overall "best practicable environmental option".[26]

Best practicable means has become the signature tune for the flexible, empirical style of pollution-control in Britain. It is a style which has survived a hundred years of scrutiny and criticism. It has been endorsed by committees of inquiry and royal commissions. It is embodied in legislation and validated by a sustained improvement in the quality of air over Britain. Of course to make such assertions is to invite the charge of complacency. The Royal Commission did not overlook some of the criticisms made in the Social Audit report which had prompted its inquiry. Nor do we overlook them. We take up this matter in our concluding chapter, but before turning to that we offer a digression in order to illustrate the technical difficulties which beset a service dedicated to the abatement of pollution.

A digression on cement

It was fortunate for the alkali inspectors that the first responsibility given to them presented no serious technical problems; muriatic acid was easy to measure and easy to remove from the gases emitted from soda works. We have discussed how parliament would not allow the inspectors to extend their control over other kinds of air pollution until practicable means for abatement were available. Thus it was that some notorious nuisances remained for many years outside the ambit of the Alkali Act because all attempts to control them failed. No example illustrates this better than the sustained struggle which the inspectors had before they could recommend a best practicable means for abating pollution from cement works. The story is worth a brief summary.

Cement is one of the most ancient building materials. It was used to line water tanks in Asia Minor in 1300 BC. The Romans discovered that a volcanic earth called *pozzolana*, mixed with lime, produced a cement resistant to fire and water. Rome, in the time of Augustus, was made chiefly of this material. In 1796 a patent was taken out in Britain for what was called Roman cement, made from a mixture of chalk and clay found on the north coast of Kent at the mouth of the River Medway. A great cement industry grew there. Coal to heat the materials was brought by sea, and the cement was carried away by sea. The best manufacturing process, invented in 1824, was to calcine the chalk and clay at a high temperature. The sintered product, called Portland cement (in the optimistic hope that it might replace Portland stone), became indispensable. Brunel used it for bridges and tunnels; it holds the bricks in the great London sewers built by Bazalgette in the 1850s.

The Portland cement industry on the shores of the Thames estuary became a money-spinner. The grey powder was carried to all parts of the world. But there was a problem: smoke from the kilns, accompanied by a nasty smell, was carried into the homes of the inhabitants of Kent and Essex. Listen to Mr Vulliamy giving evidence to the Aberdare Commission in 1877. He kept a record of his tribulations from the cement works near Greenhithe: 'Instead of...being able to walk about one's garden, and play croquet with the ladies, we were obliged to go in and shut the doors and windows.'[27] And when asked what happened when he complained, he replied: 'when you threaten proceedings, they [the cement workers] mob you, blackguard you, and throw stones at your carriage.'[28] And a partner from the cement works, in an effort to belittle Mr Vulliamy's grievance, retorted that Mr Vulliamy's nerves must be so sensitive that 'perhaps eau de Cologne would give him a headache'.[29]

Mr Vulliamy's complaints were justified. Indeed there was a much more august sufferer from cement works. It is recorded that Queen Victoria complained that ammonia from a cement works near her home at Osborne was making the royal estate uninhabitable. So the cement industry was an obvious candidate for registration under the Alkali Inspectorate. Angus Smith had said as much in his evidence to the Aberdare Commission,[30] and the Commission in its report included cement works among those which should be subjected to inspection, with a view to control when this became practical. Although opposed in principle to extending control in this way, parliament was prepared to make an exception for cement and salt works;

and the Alkali Act of 1881 included a provision that inspectors might 'from time to time enquire' whether a best practicable means could be devised for abating pollution from cement works in which aluminous deposits were used.[31] (Aluminous deposits were specified because cement made from these produced an offensive smell in addition to the smoke. Cement made from limestone marls, which contain calcium and not aluminium, made as much smoke but it was not accompanied by a smell. Cement works using marls did not come under the Act of 1881.)

But best practicable means were very slow in coming. The proposals made by Angus Smith and embodied in the Provisional Order of 1884 were rejected when brought to parliament for confirmation.[32] It was objected in committee that the means recommended would not so fully ensure removal of the nuisance to warrant their being made compulsory, and the Order was withdrawn.

Even today there is no completely satisfactory and acceptable way to eliminate the nuisance caused by cement works. Two of the nuisances, smoke and smell, have been abated by the use of rotating kilns instead of the old bottle-shaped open kilns. But the rotating kilns produced great quantities of dust. As the industry prospered and more cement works were built, the nuisance from dust became intolerable; but still the alkali inspectors were not given authority to quell the dust, for the simple reason that there were no means, practicable or otherwise, for doing it.

It was Fletcher, in many ways the most perceptive among the chief alkali inspectors, who noted, in his report for 1885, the possible application of experiments being done by Oliver Lodge, who was then professor of physics at Liverpool. Lodge showed that a high tension electric field would precipitate dust particles in the fumes from a furnace 'like flakes of snow'. In the following year Fletcher was hoping that these experiments 'on the deposition of dust by the aid of electric currents' might be applied to many manufacturing operations. He reported a test carried out in a lead works in Flintshire that year, which had failed owing partly to the primitive method of producing high tension electricity. But the idea slowly germinated, and in 1933 electrostatic precipitation was successfully applied for the first time in a cement works, the British Portland Cement Manufacturers Ltd. Johnson works in North Kent. The technique is now indispensable for power-stations, smelting works, and the like, and at long last it offered a best practicable means for the abatement of dust from cement kilns.

Promptly in 1935 it was agreed that all cement works (not just those using aluminous clays) should be scheduled.[33] Presumptive standards were fixed at 0.5 grains of dust per cubic foot. Since then standards have been stiffened: to 0.4 grains in 1946, 0.2 grains in 1962, and in 1967 to 0.2 grains for small works and 0.1 grains per cubic foot for large works. The emissions of particulates per 1000 tonnes of cement produced dropped from 16 tonnes in 1958 to 1.5 tonnes in 1974.

But even in the 1980s the abatement of cement dust bedevils the work of alkali inspectors. There are unresolved problems about the use of the equipment to precipitate the dust. The dust is carried in a very large volume of waste gases and there is a risk—since the manufacturer wants to maximize his output—that the precipitators will be overloaded; also the waste gases have a high moisture content and this causes corrosion of the electrical equipment, which has to be taken out of service from time to time. Another problem is the number of old works, on their way out, where it would be economically impracticable to insist on the installation of costly equipment; so 'with our reluctant consent' (as the Chief Inspector's report put it) some older works are allowed to exceed the presumptive standard. There was in the early 1960s a great demand for cement; and national need has to be taken into account in assessing what the best practicable means are. All this led inspectors to take a more indulgent attitude to the emissions of cement dust than was, perhaps, wise. In 1962 a freak in the weather, together with other causes, brought a record dust-fall over a densely inhabited belt in the Thames estuary. The local inhabitants were outraged. The issue was raised in an adjournment debate in the House of Commons in November 1962. The Alkali Inspectorate was put on the defensive; they pleaded that on Thames-side alone the cement industry had spent over £2 millions on dust abatement; that this area, producing over 4.5 million tonnes a year, was the most intensive area of cement production in the world; and that it had to be confessed that even at the strictest practicable standard (at that time 0.1-0.2 grains of dust per cubic foot), the area could not always be free from trouble. A further complication was that about this time the source of clay had to be changed, and this changed the conditions of manufacture.

Difficulties in abating dust from the rotating kilns were not the only trouble. There was another tiresome source of dust from the cement itself as it was stored and moved after manufacture. Some firms were having to store the cement in the open air; efforts to cover mounds of clinker with polythene sheeting were defeated by high winds. It was no wonder that the Chief Inspector, in his report for the year 1963, said that public opinion, 'roused as never before by the dust-fall of October 1962' remained uneasy and fearful of a recurrence. The modernization of cement production was not proceeding fast enough to satisfy the critics. Of 123 kilns operating, 78 had electrical precipitators, 23 had cyclonic equipment, and 22 still had no means of abatement external to the kilns. The alkali inspectors gave notice that after one more period of grace, no kiln would be allowed without external equipment efficient enough to remove all but 0.1-0.2 grains per cubic foot of dust. In 1975 they were still having to relax their standards for storing clinker 'whilst the economic recession persists'.

This digression illustrates the difficulties of administering the Alkali Act. Cement dust remains a whipping post (one of several whipping posts) to which the public like to tie the alkali inspectors. The inspectors have to strike a balance between the need for cement and the discomfort to people living near cement works. If they were to insist on a presumptive standard of (say) 0.04 grains per cubic foot—a standard which some new works have attained—then there would be less dust but more expensive cement, and some kilns would go out of business; the men who would have a grievance would be the descendants of the men who threw stones at Mr Vulliamy's carriage. It is a sobering thought that pollution from one of civilization's oldest industries still remains (in the words of the Chief Inspector) 'one of the least tractable of the many problems facing the inspectorate'.[34]

11
ASSESSMENT

The Alkali Inspectorate and Social Audit

We have already explained how the alkali inspectors have to make quasi-political decisions and how these decisions require from them skills that lie outside their competence as industrial chemists. Their tradition of autonomy and the way they used it were the burden of a critical report published in 1974 by a public interest group called Social Audit.[1]

Social Audit describes itself as 'an independent non-profit-making body concerned with improving government and corporate responsiveness to the public'. Its report, coming at the heels of several criticisms of the Inspectorate in parliament, prompted the Secretary of State for the Environment to call for an inquiry into the control of air pollution by the Royal Commission on Environmental Pollution. We discussed the Commission's findings in the last chapter.

The report by Social Audit is a lucid and on the whole accurate piece of investigative journalism. It gives credit to the Inspectorate for 'a long history [of achievements] of which it is justifiably proud' but it dislikes the style in which the achievements were made. The inspectors are criticized for the 'leisurely implementation' of the standards they set; for their 'abhorrence' of bringing offenders to court; for their 'benign policy of co-operation' with industrialists. The victims of industrial pollution, the report says, complain that the inspectors are inaccessible, remote, that they cannot even be run down in the telephone directory. The evidence, as one expects in investigative journalism, is selected to fit the theme. It is largely anecdotal and coloured here and there by innuendo. You are left to draw your own conclusions, for instance, when you are told that the alkali inspector, called in as technical assessor in an inquiry about a brickworks, was entertained to lunch by the brick company. The Social Audit asks: whose interests are served? And answers its own question: the interests of the industrialist. 'The Inspectorate's assurances that justified complaints will meet with immediate action have gathered dust in out-of-print reports.'

How is it that Social Audit finds so much wrong with the style of work of the Inspectorate and at the same time acknowledges that its success 'should perhaps be measured in terms of the number of crises it has quietly and tactfully averted'? The key to the answer lies in the Inspectorate's history. If the duty of the inspectors had been confined to enforcing statutory fixed emission limits to waste gases, it could, and probably would, have acted as the police do when they enforce a 30 mile an hour limit on a roadway. It is not for the police to relax the speed limit for motorists on urgent business or when the road is deserted. But the alkali inspectors were charged as long ago as 1874 with a much more sophisticated duty: they had to set limits themselves, using their own judgement as to best practicable means, including in this the convenience of the public and the welfare of the industry. To assume that the alkali inspector's job is simply to rid the air of noxious gases, to minimize pollution at any price, is to misunderstand his function. His job is to *optimize* pollution, which is politically a more subtle task, one that has in it an ethical dimension. The Alkali Act is administered in a way that tries to reconcile the often conflicting interests of the public who want clean air, the manufacturer who wants to make a profit, the employees who want to keep their jobs and to get a good wage, and the government who want national prosperity. Such a remit as this is incompatible with the imposition of inflexible fixed emission standards; it requires a separate value-judgement for each set of circumstances.

Since its first major revision in 1881, the Alkali Act has conferred on the Inspectorate the duty of creating the standards it has then had to enforce. This duty has deeply influenced the style of work of the inspectors. They cannot simply require compliance with rules laid down by parliament. They have to seek the co-operation of the polluters to get an agreed presumptive standard based on an acceptable best practicable means. Their relation to the manufacturer is more like that of a doctor getting the patient's co-operation in treating a disease than of a policeman apprehending a culprit. They have to win the confidence of the 'patient'. Also, like a doctor in the Health Service, they are ultimately responsible to a Minister but the Minister is not involved in diagnosis or treatment. Of course this amicable, cosy relation between the inspectors and the inspected invites the criticism that the inspector is the manufacturer's lap-dog. The report by Social Audit makes just such a criticism; it rests its case on arguments which fail to substantiate the lap-dog hypothesis. But the arguments do support other valid grounds for concern.

The Alkali Inspectorate and public relations

One ground for concern is this. The Inspectorate repeatedly emphasizes that its staff get results because they educate manufacturers; they do not—except when the manufacturer is wilfully negligent—resort to threats. (It was Angus Smith, a hundred years ago, who said that he resisted the pressure to teach 'by the cane instead of through the intellect'.) The inspectors' success in teaching industrialists is indisputable. But they act also in the interests of the public and they have an educational duty toward them too. The Inspectorate, with its tiny staff, has not felt able to do this except through occasional lectures and through its annual reports. These reports, if they are studied carefully (which most politicians and journalists do not have time to do) give a clear picture of the Inspectorate's activities, warts and all; but they make little concession in style or content to the common reader. The Inspectorate's attitude to the common reader is ambivalent. In one place a report says: 'Much as we try, it is not possible to keep entirely ahead of public opinion, *for it is public opinion which in the long run sets the standard for the community*'[2] (italics ours). In another place a report says: 'only a relatively few people are capable of properly assessing emission data' (which is undoubtedly correct) but in the same paragraph the report goes on to say: 'The Inspectorate has a statutory duty to prevent pollution from registered works and we do not want to have them diverted to the *non-productive task* [our italics] of educating numerous enquirers in the meaning of emission data'.[3] It is indeed a dilemma for the Inspectorate; the interpretation of its work to the layman is not easy. But it has been less than adept in extricating itself from the dilemma. An articulate minority of concerned citizens now demand access to facts about pollution even if they do not themselves all use the access. It is no good to say: 'what good will it do them, to know the amount of sulphur coming out of the local power station? They wouldn't be able to interpret the data anyway'. It may not do people any good to have access to the financial statements of corporations—they, too, are difficult to interpret—but there would be a public outcry if these were not available for scrutiny by anyone who cares to take the trouble.

Closely linked with this is another ground for concern: the Inspectorate's refusal to disclose information even to persons who are able to interpret it. Since 1 January 1975 inspectors have been forbidden to disclose information except to certain privileged persons, because they now work under the Health and Safety at Work Act, 1974, which has a section (28 (2)) binding them to confidentiality. But before 1 January 1975 there was no such embargo. The Act of 1906 (section 12 (2)) required inspectors to keep secret the sketch plans of the processes from which noxious gases escaped, but no other pledge of secrecy was required from them by statute. The reason for the tradition of secrecy is, of course, obvious. An inspector who failed to win the confidence of works managers would fail in his job; if he began to leak information to competitors or to the public he would lose that confidence. The inspectors' only course was to persuade industry itself to disclose data about the wastes it discharges. They have not been very successful in doing this; though one success-story, described in the Chief Inspector's report for 1970, shows how much trouble can be saved by a display of frankness. A works in the British Steel Corporation ran into serious difficulties in suppressing oxide fumes. It was not practicable to shut down the offending process. There was no hazard to health, only to amenity. The alkali inspector urged the Corporation to appeal to the public. The local authority, press, and radio were told about the trouble and the Corporation bought space in the local newspaper for a full page advertisement, telling the public why the fumes were escaping and what was being done to put matters right. The result (we quote from the Chief Alkali Inspector's report) was that 'no complaints at all were submitted and the works manager received shoals of letters thanking him for letting the public know what was happening'.[4] It is a sign of the insensitivity of some industries toward the public that this episode, which ought to have been commonplace, had to be described as a rare example of enlightenment.

Perhaps it is unfair to blame the Alkali Inspectorate because education of the public has been insufficient and insistence on confidentiality misguided. It may be argued that guidance on these matters should come from the Minister or the permanent secretary of the department. Certainly Ministers and permanent secretaries deserve some censure; if in recent years they had offered the Inspectorate a public relations officer the offer would surely have been welcomed. But it has to be remembered that the Inspectorate, until it lost some of its freedom on transfer to the Health and Safety Executive, enjoyed a privileged independence. The Chief Inspector's annual reports were signed by himself and presented direct to the Secretary of State for submission to parliament. There is no doubt that if a Chief Inspector had wanted to take the public more closely into his confidence and to disclose more information, he would have been free within the Alkali Act to do so.

One abortive attempt was made to anticipate these criticisms. The Royal Commission on Environmental Pollution published in March 1972 a brief report strongly supporting the citizen's right to know what pollutants were being put into air and water or dumped on land.[5] The Commission urged the Department of the Environment to do something to recognize this right. Following this the Clean Air Council set up a working party to examine how this could be done for industrial emissions into the atmosphere. The working party agreed with the Royal Commission and recommended that INDEMAT committees (industrial emissions to atmosphere) should be set up to provide *local* information (which is what the public want) about industrial air pollution.[6] The idea was that the INDEMAT committees would receive technical information about all kinds of air pollution in their areas and would publish reports interpreting this information to the public. The working party's report fell on stony ground and nothing came of it.

Under the Control of Pollution Act, 1974, there is another way open for information to be disclosed to the public. The Act provides for local authorities to keep records of discharges of wastes into air and water and to make these records available to anyone who cares to consult them. There are safeguards for industry: a request can be made to the Secretary of State not to release information which might disclose a trade secret. (Wastes containing traces of catalysts would come into this category.) On the ground (which we suspect is an excuse rather than a reason) that the preparation and public exhibition of these records would be costly, this part of the Act had not been implemented even six years after the Act became law.

There is another kind of disclosure which lies wholly within the discretion of the Alkali Inspectorate but which the Inspectorate shies away from making. As long ago as 1878 the Royal Commission on Noxious Vapours recommended that the Chief Inspector should publish in his annual report 'all recorded escapes with the names of the works in which they occurred'. This is not the practice. Reports of conviction for infractions of the Alkali Act are evasive. It is like putting blankets over the heads of prisoners as they are taken into court to shield them from press photographers. Firms convicted of wilful or careless pollution (and infractions have to be very wilful or careless to get as far as the courts) are protected from publicity. They appear in the annual report disguised as 'one Metal Recovery Works' or 'a Tar Works'. This gives the public what we are sure is a false impression that the Inspectorate wants to save the culprit from embarrassment. It would surely create more public confidence in the Inspectorate, and be no bad thing for industry, if offenders were to be pilloried in the comparative obscurity of the Chief Inspector's annual reports.

Another valid cause for concern—though this is not the responsibility of the Inspectorate—is the derisory level of fines imposed on offenders: £25 with 15 guineas costs was a typical figure. It is no wonder that inspectors who may have taken a lot of trouble to get evidence against offenders feel let down by the courts; let down by parliament, too, for the maximum penalty under the Alkali Act on summary conviction was £100 in 1906 and was raised (in the Schedule attached to the Control of Pollution Act 1974) only to £400.

The Alkali Inspectorate has made one serious attempt to improve its public image: it has set up, in nearly fifty places, local liaison committees. These meet about twice a year and the local alkali inspector attends to listen, to explain, and to brief his Chief Inspector. With the growing demand for participation by the public these committees are not likely to meet with approval, for their prime purpose is to inform citizens not to consult them, and they are likely to be branded as 'tokenism'. They cannot now meet the requirement in the Royal Commission's fifth report—'The public have a right to know what is in the air they breathe'—because the inspectors are now obliged to work under a statute which forbids disclosure. Still less are these liaison committees likely to be given the opportunity to be consulted before new industrial developments likely to affect the quality of the air are permitted in their areas. It will be asserted that such consultation would be a waste of time. We would deny such an assertion; there is ample evidence from Canada (to quote only one example) that time is saved and goodwill assured by judicious consultation with non-expert public-interest groups before such decisions are made. Anyone who disputes this should read the reports of the Royal Commission on Electric Power Planning in Ontario.[7]

The Royal Commission on Environmental Pollution fully vindicated the strategy of the Alkali Inspectorate. It did, nevertheless, have misgivings about the style of the Inspectorate's public relations. 'There has', wrote the Commission, 'been some clumsiness and insensitivity in the Inspectorate's public pronouncements and an air of irritation with those who presume to question the rightness of their decisions.'

The Alkali Inspectorate: misconceived criticisms

We agree with the Commission's comment, but it is only fair to the Inspectorate that we mention the extenuating circumstances. Mixed up with valid criticisms there have been some baseless charges: that the inspectors were too indulgent toward industry; that they would not explain how best practicable means operated; that they did not publish their list of presumptive standards; that they were not accountable. These accusations, all made at one time or another in parliament, must often have exasperated the Inspectorate.

Too indulgent toward industry? How, then, has it come about that works registered under the Alkali Act spent between 1958 and 1968 some £480 millions on pollution control? And why did the steel industry invest some £20 millions in capital equipment to control pollution, between 1960 and 1964, 'a massive effort with little or no financial return'; and the Central Electricity Generating Board allot £11 millions in 1965 to reduce the emissions of particulates from 0.4 to 0.2 grains per cubic foot?

Reticent about how best practicable means operates? Every issue of the annual report refers specifically to the operation of this formula. The report for 1973 includes a long essay on the history of the use of best practicable means and a Deputy Chief Inspector gave a public lecture on it in the same year.[8] When the east wind of austerity blew strongly there were many requests from industry for standards to be relaxed. This the Chief Inspector would not do. What he did agree to do was to allow a longer time for the standards to be complied with; the recession has been a time for co-operation over best practicable means, not coercion. Best practicable means, wrote the Chief Inspector, 'like some systems of Contract Bridge, require a deep understanding and lots of experience with co-operative partners'. He justified his responsibility by reminding the public that in the long run it is not the owners of works who pay for clean air, it is the public, 'and it is our duty to see that money is spent wisely on the public's behalf'.

No publication of presumptive standards? The report for 1966 describes in detail how presumptive standards are set; and in the report for 1975 there is an appendix by a Deputy Chief Inspector, M.F. Tunnicliffe, with a comprehensive list of emission limits for scheduled processes.[9]

And finally; not accountable? Accountability is being thrown as a challenge to all sorts of authorities these days. If it means that the Chief Inspector should be obliged to accept the public's recommendation on emission limits, then no, the inspectors are not accountable. But that sort of accountability, like literal participatory democracy, is a naïve way to think of making any technical decision. Desirable accountability in the public services should cover overall strategy, but not day-to-day tactics. On this understanding of accountability the alkali inspectors are accountable through parliament to the public; for the Inspectorate's report is laid before parliament and it is stuffed with factual information, statistics, and frank discussions of strategy. Anyone who reads these reports can deduce the Inspectorate's policy and through the media or through parliament can question the wisdom of that policy.

Our own verdict on these criticisms is that the prime weakness of the Alkali Inspectorate was not failure to do a good job; it was failure to communicate to politicians and to the public how good a job the inspectors were doing.

The National Survey of Air Pollution

We have already made the point that the abatement of smoke in the 1960s may have been due as much to the voluntary adoption of smokeless fuels as to enforced adoption under the Clean Air Act. Thus Norwich, Brighton, and Plymouth—three towns without statutory smoke control in the 1960s—showed nevertheless a decline in winter smoke between 1963 and 1973. It has been concluded from this example that 'none of the evidence examined lends much support for the claims made for the success of the Clean Air Act'.[10] This conclusion overlooks the fact that the very presence of the Act on the statute book stimulated the production and use of smokeless fuels, especially in industry. Evidence from four other areas testifies to the influence of the Act. London, Bristol, Coventry, and Leicester—all classified after the Beaver Report as 'black' areas in need of smoke control—had by 1975 lower concentrations of smoke in winter than salubrious towns like Ripon, Scarborough, and Whitby, which had not availed themselves of the opportunity to create smoke-control zones.

In London the change was dramatic. Most Londoners do not realize that they live in a major manufacturing city. (A few years ago it was estimated that about one-fifth of the population employed in manufacturing industries in the United Kingdom worked in London, though this proportion has fallen since then.) Before the 1950s London was a dirty city with high levels of smoke and sulphur dioxide. Today the air over London is as clean as the air over East Anglia.

People may argue about what has caused the improvement of the air over Britain but there is no argument about the evidence for the improvement itself. Fortunately it is well documented. As long ago as 1914 systematic measurements of air pollution were begun under the auspices of the Coal Smoke Abatement Society. In 1917 the work was given the blessing of government and put under the Meteorological Office. It then passed to the Department of Scientific and Industrial Research and it has been carried on since then by the Warren Spring Laboratory, at first under the Ministry of Technology and now under the Department of Industry. In 1960 it was decided to conduct a ten-year national survey of smoke, sulphur dioxide, and (in some districts only) grit and dust. Sampling stations were set up in 1122 town sites and 192 country sites. The results are published in a set of elegant volumes with maps, diagrams, and graphs which give a detailed picture of the ten-year trends at the sampling points.[11]

The survey showed that response to the Clean Air Act had been uneven, and this is still true. Some 'black' areas are as polluted as they were twenty-five years ago. But the overall improvement is impressive. Urban ground-level concentrations of smoke fell between 1958 and 1973 from about 150 micrograms per cubic metre to less than 40 micrograms. Emissions of smoke (calculated from the tonnage of soft coal consumed) fell over the same period from about 2 million tonnes to about 0.5 million tonnes. For sulphur dioxide the trend was less impressive. This is because smokeless fuels, apart from gas, contain sulphur. Emissions of sulphur dioxide remained at an approximate level of 5 - 6 million tonnes, but the urban ground level concentrations fell from around 150 micrograms per cubic metre to less than 100 micrograms. The apparent disparity between emissions and ground-level concentrations is explained by the so-called 'high chimney policy'; waste gases from power stations and other large plants are ejected at high velocity into airstreams up to a kilometre high which carry some of the gases out to sea. The one disappointing disclosure from the survey was that there had been little change in the levels of grit and dust deposited in some industrial areas.

The Clean Air Act was—as the national survey put it—'swimming with the tide of industrial development'. Many of the remaining black spots are the result of pollution from domestic fires in areas where local authorities would not, or could not, bring in orders to control smoke. Some authorities, as we have mentioned already, are constrained by the desire of miners to keep their right to an allowance of concessionary coal. The National Coal Board has done its best to compromise over this desire: miners have been offered cash sums in lieu of the concession, or allowances of smokeless fuel. The hope expressed in the Beaver Report, that by about 1970 all 'black' areas would be covered by smoke-control orders, has not been fulfilled. But the achievement is promising. Some 7 million premises are covered by orders, representing nearly 60 per cent of the target set by the local authorities themselves.

The legislative arrangements for cleaning Britain's air to an acceptable level are sufficient, and public opinion is well disposed toward measures to abate pollution. The principal constraints are technical and financial. Steady work is going on to tackle the technical problems. The Department of the Environment has sponsored working parties on grit and dust emissions[12] and on the suppression of odours from offensive trades.[13] It has issued a series of Pollution Papers on various aspects of control.[14] A good deal of monitoring goes on, conducted by local authorities, the alkali inspectors, special research groups, and industry itself.[15] As to the financial constraints, the problem in the present economic climate is to make sure that the limited resources available to protect the environment are spent in the most effective way. We discuss the cost–benefit analysis of pollution abatement below. But first we have a brief digression to make.

A digression on pollution from automobiles

We shall doubtless be criticized for paying so little attention to the air pollutants which some people regard as the most menacing of all, namely emissions from the exhausts of automobiles. Our defence is that this book is about the politics of clean air, and agitation about car exhausts did not arouse official concern until the 1970s, too recently to be assessed by the historian. Under the Motor Vehicle (Construction and Use) Regulations of 1966 it is an offence for a car to emit 'avoidable smoke or visible vapour'—a proscription which applies mainly to diesel engines. It is only very recently in Britain that public anxiety is compelling the government to prescribe limits for emissions of lead, carbon monoxide, and oxides of nitrogen.

We have one comment to make about this anxiety of the public. In Britain there is no evidence that any of the emissions from automobiles are as serious a hazard to health as are the emissions from domestic fires burning soft coal. The amount of carbon monoxide you inhale in a busy city street rarely reaches the amount inhaled after smoking three or four cigarettes; nor does carbon monoxide accumulate in the atmosphere, as carbon dioxide does: it is consumed as fast as it is formed by benevolent microorganisms rejoicing in the names *Methanosarcina* and *Carboxydomonas.* As for lead, the amount you inhale from car exhausts, though under suspicion that it might have long-term sub-clinical effects on children, is only about a quarter of the amount you absorb from other sources.

In the USA more is spent on the abatement of pollution from automobiles than upon any other kind of pollution. This may be appropriate in the USA but it would not be appropriate in Britain. For our circumstances, a pound spent on the abatement of smoke and sulphur dioxide in the remaining 'black' areas would be a better investment in the nation's health than a pound spent on the abatement of carbon monoxide or lead from automobiles. With limited resources, what matters is cost-effectiveness.

Clean air: costs and benefits

On p. 140 we gave a few examples of the costs incurred by some industries in order to comply with the standards set by the Alkali Inspectorate. The costs of abating smoke in order to comply with orders under the Clean Air Act are much more elusive. Much of the change-over from soft coal to smokeless fuels has been as automatic as the change-over from carpet sweepers to vacuum cleaners, and has nothing to do with legislation. It is possible, however, to get a very rough idea of costs from the sums spent by local authorities on subsidies to householders who have been obliged to convert to smokeless fuels. The householders receive 70 per cent of the cost of conversion (four-sevenths of which is provided from the central government), and they have to pay the balance of 30 per cent themselves. In 1970, when just over half the premises in black areas were covered by smoke orders, the total cost of the conversions, including the householder's share, came to something like £40 millions, at 1970 prices.

The Americans have gone to much more trouble than we have to estimate costs of abating air pollution. The Council on Environmental Quality publishes these estimates in its annual reports. Figures for 1978 are given in the Council's tenth annual report.[16] Excluding the costs of abatement of emissions from automobiles, the estimates for other kinds of abatement of air pollution came to $11.2 billions. The most interesting figure is the estimate for incremental costs, arising from federal legislation, and over and above what would have been spent in the absence of that legislation. This estimate came to $9.0 billions, that is, 77 per cent of the total expenditure. Even though the estimate is open to the same criticism as similar estimates are open to in Britain, namely that Americans, like the British, are voluntarily abating pollution anyway, it is an impressive figure.

Society does not willingly pay costs as heavy as these (except for armaments) without some hard evidence that the costs are matched by benefits. So the Americans have gone to great trouble to try to quantify the benefits of clean air. The findings are summarized in the annual reports of the Council on Environmental Quality. A diagram in the ninth report compares the expected damage which will be done by air pollution up to 1985 with the damage that would be done if there were no control of air pollution.[17] The difference for 1980 (and hence the benefit arising from pollution control) is of the order of $36 billions. If the assumptions that went into this calculation were acceptable (a point on which we have some doubt) it would convince an American that he was getting an excellent bargain in the money his country was spending on the control of pollution.

In Britain there has been one heroic effort to quantify the damage done by air pollution. In the surge of enthusiasm for abating pollution toward the end of the 1960s, politicians were driven to ask how much the nation could afford to spend on cleaning up air and rivers and coastal waters; and this led to another question: how clean is clean enough? There were no analyses on which to base answers to these questions. Accordingly the Ministry of Technology and the Ministry of Housing and Local Government jointly sponsored a study by a group attached to the Department of Industry and the Atomic Energy Authority, called the Programmes Analysis Unit. The Unit delivered their report to the Ministry in 1970. The Ministry was coy about releasing the study: it was given a limited circulation in 1971, and an edited edition was published in 1972 in mimeograph form.[18]

After ten years of inflation the arithmetic of the study is of course out of date. But comprehension of the method used and the assumptions made for the calculations is still essential to any study of the politics of clean air. We shall criticize the way some of the assumptions are used to support political decisions, so we want to make it clear that the prime purpose of the study was to assess the relevance of current programmes of research and development, not to offer guidelines for legislation.

We called this study by the Programmes Analysis Unit a 'heroic effort'. Why? Anyone doing a cost–benefit analysis of measures to abate pollution has two options. The easier option is to include only the simple quantifiable economic benefits and to put the onus on the politician to take account of unquantifiable social benefits. The harder option is to assert that all benefits—economic or social—can be quantified in terms of money, and to give the politician a balance sheet of costs of abatement in one column alongside benefits from abatement in another column. The Programmes Analysis Unit boldly chose this second option, though they conscientiously and repeatedly warned the reader about the pitfalls in this choice.

Put very simply, the method of analysis was as follows. Britain can be very roughly classified into polluted areas, (the 'black' areas which the Beaver Committee wrote about) and 'clean' areas. Using as a criterion the prevalence of high levels of sulphur dioxide in the air, the Unit concluded that the polluted areas of Britain contained some 6 million dwellings and the 'clean' areas some 12 million dwellings. Smoke and sulphur dioxide in the air damage stonework, textiles, and paintwork. They affect property values; they corrode metals; they may lower the yield of crops and they may be a hazard to health. These effects may be negligible in remote rural areas; they are severe in the main conurbations. So how much more do the occupants of the 6 million dwellings in polluted areas have to pay, in economic and social disbenefits, compared with the occupants of the 12 million dwellings in clean areas? A credible answer to this question (if one could get it) would be a rough-and-ready guide to the marginal benefits to be expected if the polluted areas were to be made as clean as the clean areas. This figure for all the marginal benefits could then be set against the marginal costs of cleaning up the polluted areas. With the help of some ingenious assumptions, the Unit examined one by one the marginal costs attributable to air pollution. We shall have to content ourselves with two examples.

The first example: property. Anyone who moved in the 1950s into one of the 'black' conurbations soon discovered that shirts got dirty more quickly, windows had to be cleaned more often, paintwork decayed, the family car had to be washed more regularly if the owner had any pride in its appearance. The Unit tried to quantify these additional costs from evidence about sales of soap, costs of laundry, frequency of washing windows, and the like. Thus, they found that ground floor windows in polluted industrial areas were being cleaned about once every three weeks, while in non-industrial towns they were cleaned only about twice in every three months: a calculation which gave an additional cost of window cleaning in polluted areas of the order of £5 millions a year. Similarly, if stonework in polluted areas was to be kept as clean as stonework in clean areas, the additional cost would be of the order of £1 million a year. But, as the Unit discovered, if you live in a 'black' area, you (meaning the average citizen) do not keep your laundry, your paintwork, your car, your town hall, as clean as citizens do who live in 'clean' areas. So a great deal of the additional cost of preserving property in these two kinds of environment is never incurred. The Unit found that the 'hard' economic costs of keeping things clean turned out to be no more than £5.5 millions a year—most of this sum was for cleaning windows—simply because other things were not kept so clean. But the 'soft' social costs of, so to speak, cleanliness forgone, was reckoned to be £166 millions a year, of which 90 per cent (£150 millions) was an estimate of the social costs of the additional labour of housewives to keep their clothes, curtains, and sheets clean! The moral we draw from this example is that the disamenities of life in a heavily polluted place, vivid enough to anyone dwelling there, cannot be made more credible by putting a money tag on them.

The second example: health. The epidemiological evidence that smoke and sulphur dixoide are hazards to health was summarized in a report issued by the Royal College of Physicians in 1970.[19] These two pollutants are much more dangerous than the pollutants that escape from the exhausts of cars. This assertion is supported by the familiar evidence, that fogs in the days before the Clean Air Act killed sufferers from respiratory diseases. Even more significant is evidence of the long-term effects of polluted air upon children. In one survey, analysed by Douglas and Waller, 3866 children born in 1946 were periodically examined by health visitors and school doctors.[20] It was found that 12.9 per cent of these children living in highly polluted areas had repeated attacks of bronchitis; in areas of low pollution only 4.3 per cent of the children suffered attacks. It is, of course, difficult to disentangle the effects of air pollution from other possible causes of ill health, for children who live in areas of high pollution may be less privileged socially than other children and may be less well nourished. The report had to admit that 'there is as yet no conclusive proof that air pollution in particular, rather than urban conditions in general, predisposes to infections of ear, nose, and throat in children'. Nevertheless the balance of evidence does indict air pollution as a serious hazard to child health.

The Programmes Analysis Unit, after weighing the evidence, came to the conclusion that 'it would seem wise to assume that air pollution could be responsible for excess bronchitis in polluted areas of anywhere from 0 to 50% with a median value of 25%'. Having opted for quantification, the Unit then set out to put a money value on the damage to health caused by pollution. With renewed warning to the reader, the Unit estimated separately the economic and social costs, as they had for household property. Chronic illness or premature death deprive society of a productive worker. The cost of this deprivation, together with the cost to society for medical treatment of the invalid, (with various subtle qualifications) constitute the marginal economic damage attributable to the polluted environment in which the 6 million citizens live. As to the social cost, on the option chosen by the Unit a money tag had to be attached to the suffering and perhaps bereavement in a family stricken by illness caused by air pollution. There is experience bearing on this in other fields: compensation for industrial injuries, awards of damages after road accidents. The Unit considered the various ways in which a value can be attached to a human life: What compensation do courts award for deaths caused by industrial accidents? What sums will society (or government acting on behalf of society) pay to save a 'statistical' life; e.g. by putting crash barriers on a road or fire doors in hotels? What will people pay to regain health or to avoid future ill health, e.g. by moving at some financial loss from an unhealthy to a healthy place of work?

To pursue these questions would carry us into the thorny topic of risk-benefit analysis. We shall not make this pursuit because there is no shortage of books on this topic.[21] It is the overall conclusion which bears upon our story, namely that there is no predictable or rational relation between the statistical assessment of a risk and the ways in which individuals perceive a risk. It follows that the sort of rational analysis which the Programmes Analysis Unit used, while it may well throw light on the relevance of research and development, is of little use as a guide for political decisions, for these have to take into account perceived rather than statistically assessed risk.

As for property, so for health, the Unit separated the hard economic costs of ill health from the 'soft' social costs; and the outcome was a similar one: the marginal costs which would be saved by an abatement of pollution for bronchitic patients comprise an economic component (loss of production, medical treatment) and a social component (the disamenity due to illness, the bereavement following death of the patient). Over 70 per cent of these costs are speculative money tags put upon the social disamenities of having chronic bronchitis.

The Programmes Analysis Unit were at great pains to emphasize that the reader should not take these figures too seriously. Nevertheless this kind of analysis does offer the politician a simple figure (e.g. in this case a benefit amounting to some £62 millions in relief from bronchitis if pollution is abated), which the politician is naturally tempted to use as a baseline for making policy. Quantification carried to this extreme is harmless in the hands of experienced economists who are used to the symbolism of shadow pricing. But in the hands of politicians and other decision-makers it can be dangerous. It conceals the fact that there is a great deal more precision behind estimates of the cost of abating pollution than there is behind estimates of the damage done or the benefits forgone by not abating pollution.

It is on grounds such as these that we come to the conclusion that cost-benefit analysis is an unreliable tool for fashioning policies for the environment. To be able to decide whether one way to abate pollution is more cost effective than another is, of course, very important; but the decision whether or not to abate pollution cannot, in our view, be made by measuring the cost of abatement against some quantified version of the benefits which will follow. To quantify fragile values (the tranquillity of a river, the clarity of the atmosphere, the relief from bronchitis, freedom from the drudgery of constant laundry-work) is to drain them of their meaning. Cost–benefit analysis asks: what is efficient for society. When a political decision has to be made about protection of the environment this is not the right question. What is at issue in the politics of clean air is not simply: What is efficient for society; it is what is good for society. This question has a moral dimension that lies beyond the range of economics but does lie at the centre of politics.[22]

The international dimension: acid rain

The theme of our book is the evolution of a national policy for clean air over the last 160 years. But there has to be an international policy too: some kinds of pollution have no respect for frontiers, and Britain has recently realized, to her embarrassment, that she is exporting pollution to some of her friendly neighbours.

Between 1956 and 1965 the acidity of rain over parts of Scandinavia doubled, owing to the presence of small quantities of sulphuric and nitric acids in the rainfall.[23] This caused concern, particularly in Sweden, because it was feared that the acid rain was damaging crops and fisheries. The Swedes prepared a special report on this matter which was presented at the United Nations conference on the environment in Stockholm in 1972.[24] The report asserted that the increase in acidity of the rain was caused by pollutants, especially sulphur, carried into Sweden from the burning of coal and oil in neighbouring countries.The acid rain was blamed for increased corrosion of metals and deterioration of stonework, lower productivity in forests, and hazards to fish. There was even an estimate of the financial loss arising from the higher acidity of the rain; a summary of it is given in Table 2.

Table 2.

Cause of expense or loss	Loss in million kroner p.a.
Liming of fresh water	4.8
Liming of agricultural land	4.7
Loss from reduced forest growth	73.5
Reduced production of cranberries, etc.	4.0
Estimated total annual loss due to acid rain	87.0

These assertions, if they are correct, amount to a serious charge of unneighbourliness against the nations exporting pollution to Scandinavia. Are they correct? And if so, what is Britain's share in making the nuisance?

In 1972 the Organization for Economic Co-operation and Development (OECD) undertook a study of the long-range transport of air pollutants.[25] Eleven European nations took part. The purpose was to decide how much pollution from sulphur in the air was generated within the co-operating nations, how much was imported from other nations, and how much was exported. Seventy-six sampling stations were set up, in rural areas so that the records would not be affected by local emissions. Concentrations of sulphur and compounds formed from sulphur were measured also from aeroplanes over the North Sea. The interpretation of the data was a ticklish exercise. With the aid of mathematical models of air flow, a very rough balance sheet for the deposition of sulphur over Europe was drawn up. The figures are accurate only to within plus or minus 50 per cent. Nevertheless, even allowing for this broad band of error, only the most captious critic would dispute the implications for Britain. These are that in 1973–74 (the period chosen for the sampling) Britain put into the air between five and six million tonnes of sulphur dioxide a year. Sixty per cent of this left the country, to be carried several hundred kilometres by the prevailing winds. There is an excess of sulphur in the air over and above the amount generated from the burning of fuels in Europe. Some of this comes from natural sources and some may come across the Atlantic from America. The most interesting figure—and the most dangerous one to quote without the reservations attached to it—is the contribution made by Britain to sulphur-pollution in Sweden (and other Scandinavian countries). Of the nations known to contribute to sulphur-pollution in Sweden, Britain's share is estimated at 16 per cent, or, including the sources unaccounted for, 11 per cent. On the same assumptions Britain's contribution to the budget of sulphur-pollution in Norway is about 47 per cent of the importation from known sources, and 26 per cent from all sources. The principal other nations contributing sulphur-pollutants to Scandinavia are Western Germany and the German Democratic Republic.

The evidence that the imported pollution is harming agriculture and fisheries in Sweden rests on less secure foundation. The productivity of forests is affected by other environmental and climatic factors; these effects have not yet been disentangled from the possible effects of acid rain. As for damage to fisheries, the salmon catches in some Swedish rivers were falling long before acid rain became a problem. The possible effects of acid rain have still to be disentangled from changing practices of land-use (affecting run-off into rivers), epidemic disease, and over-fishing.[26]

However, after all these displays of scepticism, the uncomfortable fact remains that something like three million tonnes of sulphur dioxide were carried away from Britain in 1974, to be deposited somewhere else, and until our output of sulphur dioxide falls, this will continue. The environment of Britain is not suffering seriously from excess of sulphur dioxide from industry (though there is room for much improvement in the remaining 'black' areas); so as a national policy, the de-sulphurization of coal and oil does not have a high priority. But should it have a high priority as an international policy? The Scandinavians think it should and they are sharply critical of British policy. But the hard facts are that the diseconomy suffered by Sweden—only a few million pounds even on Sweden's own estimate—is trivial compared with the enormous cost of ridding coal (and to a less extent oil) of sulphur. The Central Electricity Board has estimated that the cost to coal-fired power stations alone of installing flue gas desulphurization plant would be of the order of £2 billions, and the running costs of the order of £500 millions a year. The journal *Nature* summed up the position in the headline to its editorial commenting on the OECD report: 'Million dollar problem—billion dollar solution?'

The international dimension: the European Community

Britain's membership of the European Community brings an even more compelling international dimension into the politics of clean air. British practice now has to be consistent with practice in other member States.

This, too, is causing Britain embarrassment, for British policy differs from the policy which the European Commission is trying to impose on member States.

The legal status of the Community rests upon the Treaty of Rome. There is no reference to the environment in this Treaty. Directives about the abatement of pollution rest their legitimacy on Article 100 or Article 235 of the Treaty.[27] Article 100 provides for an approximation of laws among member States so that trade between the States is not distorted; but many environmental directives have no relevance whatever to trade. Article 235 empowers the Community to take measures to deal with any unforseen circumstances which would otherise impede the objectives of the Community; but protection of the environment is not one of the formal objectives of the Community.

On this ground, legalistic persons regard the whole environmental policy of the Community as *ultra vires*. But if the Community is ever to be more than a consortium for trade there will have to be common objectives which lie outside a narrow interpretation of the Treaty; and it is on this assumption that the Community has a policy for the environment. The policy arose in this way. In 1972 there was a meeting of heads of States in the Community (before Britain had joined). A statement issued from this meeting included the following passage:

> economic expansion is not an end in itself: its first aim should be to enable disparities in living conditions to be reduced...It should result in an improvement in the quality of life as well as in standards of living. As befits the genius of Europe, particular attention will be given to intangible values and to protecting the environment so that progress may really be put at the service of mankind.

To this end the heads of state invited the Community institutions to establish a Community environmental policy. In response to this invitation the Council of the European Communities produced a Declaration in November 1973 which remains the blueprint for the Community's programme.[28] It is in fact the only authority for the Community to have an environmental policy at all. There is nothing in this Declaration inconsistent with British policy for the environment. It contains two important provisos: one, that the protection of the environment should be conducted 'at the lowest cost to the community'; the other, that in setting quality objectives for the environment 'proper account must be taken of the specific characteristics of the regions in question'. These provisos allow just that degree of flexibility which is characteristic of the pragmatic approach developed by the Alkali Inspectorate and the permissive arrangements for decisions about smoke-control orders under the Clean Air Act.

But this is not the way the European Commission in Brussels is interpreting the 1973 Declaration. It has, so far, been the practice of the Commission to issue directives to harmonize effluent standards into rivers and emission standards into the air without regard to local conditions and often without regard to keeping down the cost to the community. It is a policy not of harmony in the musical sense but of unison; uniformity from the north of Scotland to the south of Greece. If some of these directives were to be adopted for Britain there would be massive needless expense and no room for the discretion allowed under the formula of best practicable means.

Most of the controversial directives already issued to protect the environment refer to the abatement of pollution in water, not in air. But here is one example of this policy of harmonization applied to air pollution, and more are likely to follow. It is in a directive 'relating to the use of fuel oils with the aim of decreasing sulphurous emissions'.[29] Article 2 of this directive prescribes that member states shall use only low sulphur fuel oil in regions where the median daily average concentration of sulphur dioxide over a year exceeds 80-120 micrograms per cubic metre (the lower level of 80 when the median annual average of particulate matter exceeds 40 micrograms per cubic metre, and the higher level of 120 when the level of particulate matter is less than 40 micrograms per cubic metre.) The choice of these arbitrary levels of sulphur dioxide and particulate matter is disputed by the British experts, but that is not the main point; what is more serious is the rigidity of the directive. For some member states it makes sense, for others it does not; and the reason for the difference is simple. Generating stations for electricity are the main source of pollution from the burning of oil. The other fossil fuel burnt at generating stations is coal, which carries a high sulphur content. In Belgium, Ireland, and Italy, less than 20 per cent of electricity comes from coal-fired power-stations and in these countries over 50 per cent of electricity comes from oil-fired power-stations. Therefore in these countries the use of low sulphur oil would be likely to make a significant reduction in the level of sulphur dioxide. But in Britain over 60 per cent of electricity comes from coal-fired power-stations and only about 26 per cent from oil-fired stations. Therefore the obligatory use of low sulphur oil in Britain, except in certain very limited areas (the City of London is one of them) would simply be a waste of resources. If it were practicable to de-sulphurize coal at a reasonable cost, the situation would be different; but in present circumstances this is out of the question.

There is, therefore, a basic difference between the environmental policies of Britain and the European Community. It is a difference, we think, in jurisprudence. In Britain we are accustomed to a legal system based on common law, which grows by a gradual accretion of precedents. Other nations in the Community have a different legal system which grows by successive decrees from above. There is a Cartesian elegance about the tidy, comprehensive legislation which aims to harmonize policies for clean air and water over the length and breadth of Europe. British legislation must appear to our fellow Europeans to be pragmatic, piecemeal, *ad hoc*, the product of expedience, not principle: a policy to be described as a non-policy. Yet British policy has deep roots in history. It is the product of nearly two centuries of evolution in which impracticable ideas have been eliminated, Utopian aspirations have been discarded, and the policies which have survived have been proved to work. It was Churchill who is reputed to have said: 'The English never draw a line without blurring it.' There are hopeful signs that we may find ways to reconcile the British and the European policies for managing the environment; but it would be a great mistake if the price for reconciliation were to be a repudiation of the lessons we have learnt from 160 years of our own history.

APPENDIX

The Leblanc Process for Alkali-Manufacture

The story of the Alkali Inspectorate could have been written on a different theme, to bring out the scientific and technical problems handled by the Inspectorate. We have not dealt with this theme because our book is addressed to the general reader, and the scientific and technical theme has been developed in several books on the history of the chemical industry.[1] There is, however, one industrial process which we must describe in simple terms for the non-technical reader, namely the process which gave birth to the Alkali Inspectorate.

Alkalis containing sodium are used in the manufacture of soap, glass, textiles, paper, and in some metallurgical processes. They were therefore in great demand in the early days of the chemical industry. In 1787 a Frenchman, Nicolas Leblanc, devised a process for making soda-alkalis from common salt. The process was introduced into Britain by James Muspratt in the 1820s at St Helens, near Liverpool, where the essential ingredients—salt, limestone, and coal—were handy. Soon afterwards, in 1825, an alkali works was set up in Glasgow. Trade flourished, and in the 1830s the works in Glasgow was the biggest in Europe. It covered 100 acres and had some thousand employees.

The first stage in the process was the making of *salt cake*. Common salt (sodium chloride) was mixed in a shallow pan with sulphuric acid, stirred, and gently heated. An exchange of elements took place, to produce an intermediate substance (sodium hydrogen sulphate) and the notorious hydrochloric acid (called muriatic acid), which escaped as a gas. The intermediate substance was then pushed with a rake on to the bed of a furnace and roasted at high temperature. This converted the intermediate substance into salt cake (sodium sulphate), with a further discharge of hydrochloric acid.

The pollutants put into the air were hydrochloric acid from both steps in the process and a good deal of sulphurous smoke from the furnace which roasted the salt cake.

The second stage in the process was the making of *black ash*. Salt cake is a hard yellowish solid. It was broken into lumps, mixed with coal and limestone, and shovelled into a large drum which rotated as flames from a furnace passed through it. The flames ignited the coal mixed with the salt cake. Carbon from the coal replaced sulphur attached to the sodium. The chunks of solid material in the cylinder became a sticky mass under this treatment. At the right moment the drum was opened and the contents were run out into iron trucks, where they solidified into a porus grey substance, called black ash. The chemical change was to turn the salt cake into a mixture of sodium carbonate (washing soda), which was the desired alkali product, and a nasty waste material, consisting largely of calcium sulphide. Its nastiness can be imagined from the trade name given to it: it was called *galligu*.

Sodium carbonate dissolves in water; calcium sulphide does not. So by repeated washing the desired alkali was leached out of the mixture. The liquid, called *tank-liquor*, still contained some impurities.

The pollutants produced in this stage were sulphurous smoke from the rotating furnace, and the solid galligu.

The third stage in the process was to purify the tank-liquor. It had to be treated to remove caustic soda and other contaminants. It was then once more roasted in a rotating furnace and the commercial product, called *finished soda ash*, was ready for the market.[2]

The Alkali Act of 1863 set an emission limit for hydrochloric acid, but there was no control over other pollutants in the process: the sulphurous smoke and the noisome galligu. Accumulations of galligu became a serious problem, for two tons of it remained for every ton of soda manufactured. Later on it was discovered that galligu itself could be a source of profit. If it is heated in limestone kilns it produces a gas, sulphuretted hydrogen (hydrogen sulphide) from which sulphur—a valuable product in those days—could be recovered.[3] But this gave rise to a fresh nuisance, for sulphuretted hydrogen is poisonous and smells like bad eggs. That is why alkali inspectors pressed so hard for this process to be brought under their control.

In 1872 Ludwig Mond acquired the rights in a more efficient and less polluting process for making soda, known as the ammonia-soda process. But owners of works using the Leblanc process, having sunk a lot of capital in their equipment, were unwilling to change their practice, and it was not until the early twentieth century that the Leblanc process was phased out. There was a drawback, too, to the ammonia-soda process: it did not allow an easy recovery of chlorine, which had become a valuable by-product because out of it bleaching powder could be made.

So in the long run, the two unacceptable waste products of the Leblanc process turned out to be benefits: bleaching powder could be made from the hydrochloric acid and sulphur could be recovered from galligu.

Notes

1. Clow, A. and Clow, N.L., *The chemical revolution: a contribution to social technology*. Batchworth Press, London (1952).
   Haber, L.F., *The chemical industry during the nineteenth century*. Oxford University Press, London (1958).
   Hardie, D.W.F., *A history of the chemical industry in Widnes*. Imperial Chemical Industries, London (1950).

2. In outline the stages are:

(1) $NaCl + H_2SO_4 = NaHSO_4 + HCl$

(2) $NaHSO_4 + NaCl = Na_2SO_4 + HCl$

(3) (a) $Na_2SO_4 + 2C = Na_2S + 2\,CO_2$

(b) $Na_2S + CaCO_3 = CaS + Na_2CO_3$

3. The recovery of sulphur is as follows:

$$CaS + H_2O + CO_2 = Ca\,CO_3 + H_2S$$

$$2H_2S + O_2 = 2H_2O + 2S$$

REFERENCES

Abbreviations

AR Annual report of the proceedings of the Alkali Inspectorate (See Chapter 2, ref. 16. Following the creation of a Secretary for Scotland in 1887, the Chief Alkali Inspector was required to make a separate report for Scotland. These are referred to as AR (Scotland))
HLG Ministry of Housing and Local Government records (Public Record Office, London)
HO Home Office records (Public Record Office, London)
MH Ministry of Health records (Public Records Office, London)

Chapter 1

1. *The Times*, 17 Feb. 1881, 9e.
2. *HC Debs*, 8 June 1819, col. 976.
3. *Parl. Pp.* (*HC*) 1819 (574) VIII.
4 Ibid., 1820 (244) II.
5. 1 and 2 Geo. IV cap. XLI.
6. *HC Debs*, 2 May 1820, col. 50.
7. Ministry of Health, Committee on smoke and noxious vapours abatement, *Final report*. HMSO, London (1921) 35 [Newton Report].
8. Evidence to the Mackinnon Committee, 1843; *Parl. Pp.* (*HC*) 1843 (583) VII, q. 1809.
9. Ibid.
10. Evidence to Taylor's committee, 30 May 1820; ref. 4.
11. Statement by M.A. Taylor; ibid.
12. Evidence, 5 July 1820; ibid.
13. *HC Debs*, 2 May 1820, col. 52.
14. Ibid., 18 Apr. 1821, col. 439; *Parl. Pp.* (*HC*) 1821 (434) II.
15. *HC Debs*, 7 May 1821, col. 535.
16. An Act for giving greater facility in the prosecution and abatement of nuisance arising from furnaces used in the working of steam engines; ref. 5.
17. Ref. 8, q. 1850.
18. Cf. Mackinnon, W.A., *History of civilisation*. Longmans, London (1846).
19. 6 Geo. IV cap. CXXXII, sect. 65.
20. 5 & 6 Vict. cap. CIV, sect. 249.
21. *HC Debs*, 2 May 1820, col. 51.
22. *HC Journal*, 13 June 1843.
23. *HC Debs*, 27 June 1843, col. 445.
24. Ref. 18.
25. Ibid.
26. *Parl. Pp.* (*HC*) 1843 (583) VII.
27. 9 May and 11 July 1845; ibid., 1845 (289) XIII, (489) XIII.
28. Ref. 26, q. 1476.
29. Ibid., q. 1992.
30. Ibid., qs 1856-63.

31. Ibid., q. 1816.
32. Ibid., q. 156.
33. Ibid., q. 2062.
34. *Parl. Pp.* (*HC*) 1844 (275) IV.
35. Ibid., 1845 (602) XVIII.
36. Ibid., 1845 (94) VI.
37. *HC Debs*, 12 Mar. 1845, col. 727.
38. *Parl. Pp.* (*HC*) 1845 (489) XIII, q. 940.
39. *The Times*, 24 Apr. 1845, 4f.
40. *Parl. Pp.* (*HC*) 1846 (194) XLIII.
41. *HC Debs*, 1 May 1846, col. 1352.
42. *Parl. Pp.* (*HC*) 1846 (371) IV.
43. *HC Debs*, 12 Aug. 1846, col. 629.
44. *Parl. Pp.* (*HC*) 1847-48 (83) V, cl. 42.
45. *Parl. Pp.* (*HL*) 1847-48 (102) VI.
46. *HC Debs*, 7 Aug. 1848, col. 1178.
47. *Parl. Pp.* (*HL*) 1849 (81) IV; 1850 (106) VI.
48. 10 & 11 Vict. cap. XXXIV sect. 108.
49. *HC Debs*, 11 July 1849, col. 195.
50. Ref. 8, q. 55.
51. Simon, J., *Reports relating to the sanitary conditions of the City of London.* J.W. Parker & Son, London (1854).
52. Lambert, R., *Sir John Simon, 1816-1904 and English social administration.* Macgibbon & Kee, London (1963); ref. 51.
53. *HC Journal*, 15 June 1852.
54. HO 45/4761.
55. *Parl. Pp.* (*HC*) 1852-53 (829) VI.
56. *HC Debs*, 8 Aug. 1853, col 1495 ff.
57. Ibid., col. 1496.
58. 16 & 17 Vict. cap. CXXVIII.
59. 5 Sep. 1853; HO 45/4761.
60. 16 July 1854; HO 45/5677.
61. 4 Dec. 1853; HO 45/4761.
62. *The Times*, 12 Aug. 1854, 8f.
63. 14 Aug. 1854; HO 45/5677.
64. 15 Aug. 1854; ibid.
65. 28 Aug. 1854; ibid.
66. 2 Sep. 1854; ibid.
67. Metropolitan Police Office to HO, 11 Sep. 1854; ibid.
68. Treasury to HO, 16 Nov. 1854; minutes by Palmerston, 7, 11 Dec. 1854; ibid.
69. Return to address of HC, 12 Mar. 1855; *Parl. Pp.* (*HC*) 1854-55 (270) LIII.
70. General Board of Health to HO, 25 Nov. 1853; HO 45/4761.

Chapter 2

1. Cf. Derby Report, ref. 4 below.
2. *The Times*, 12 May 1862, 8e.
3. *HL Debs*, 9 May 1862, col. 1452.
4. *Parl. Pp.* (*HC*) 1862 (486) XIV.
5. Ibid., q. 1067 ff.
6. Ibid., q. 1537.

7. Ref. 4, q. 2336.
8. 26 & 27 Vict. c. 124.
9. Ibid., sect. 19.
10. To Treasury, 2 Sep. 1863, enclosed in memorial to Local Govt Bd, Sep. 1874; MH 16/1.
11. To Treasury, 22 Sep. 1863; ibid.
12. Quoted MacLeod, R.M., 'The Alkali Acts administration, 1863-84...', *Victorian Studies*, IX, No. 2 85-112 (Dec. 1965).
13. F.H.S., 'Robert Angus Smith', *Am. J. Sci.*, 3rd ser., XXVIII 79-80 (1884).
14. Smith, W. Anderson, '*Shepherd' Smith the universalist*... p. 17. London (1892). For biographical details of R.A. Smith, see also Gibson, A., *Robert Angus Smith and sanitary science.* Unpub. M.Sc. thesis, University of Manchester Inst. of Science and Technology (1972).
15. To Secy, Local Govt Bd, 8 Dec. 1880; MH 16/1.
16. It was a requirement of the Alkali Act, 1863, that the Inspector should, on or before the first day of March each year, make a written report to the Board of Trade of his proceedings during the previous year, and that a copy of this report should be laid before both houses of parliament. Smith presented his first report in Feb. 1865 (*Parl. Pp.* (*HC*) 1865 (3640) XX). This and subsequent reports will be referred to as 'AR for...' followed by the date of presentation to parliament, and, when there was a delay in publication, the date of completion.
17. *AR for 1864* (1865) 40-1.
18. Ibid., 19-20.
19. *AR for 1873* (1874) 406.
20. Ref. 17, p. 36.
21. Ibid., p. 61.
22. *AR for 1865* (1866) 8.
23. *AR for 1866* (1867) 53.
24. *AR for 1868* (1869) 30.
25. *AR for 1872* (1873) 5.
26. Ref. 15.
27. *AR for 1875 and 1876*, Feb. 1877 (1878) 26.
28. *AR for 1881*, Aug. 1882 (1883) 8.
29. *AR for 1864* (1865) 40.
30. *HL Debs*, 22 May 1865, cols 628, 631.
31. Alkali Act, 1863, Perpetuation, 31 & 32 Vict. c. 36.
32. *HC Debs*, 9 June 1868, col. 1296.
33. Ibid.
34. *AR for 1871* (1872) 5.
35. *AR for 1873* (1874).
36. 'Chemical nuisance legislation', official memorandum, 22 June 1872, submitted in evidence to the Aberdare Commission, 6 June 1877; *Parl. Pp.* (*HC*) 1878 (C 2159-I) XLIV, q. 13180.
37. *AR for 1867* (1868) 89.

Chapter 3

1. Alkali Act (1863) Amendment Bill; *Parl. Pp.* (*HC*) 1874 (99) I.
2. *HL Debs*, 25 June 1874, col. 386.
3. Ibid., 2 July 1874, col. 863.
4. 37 & 38 Vict. c. 43.

5. To Secy, Local Govt Bd, 24 July 1875; MH 16/1.
6. Interim report of proceedings since passing of Alkali Act, 1874, 9 Nov. 1875; *Parl. Pp.* (*HC*) 1876 (165) XVI.
7. *HL Debs*, 24 Feb. 1876, col. 790.
8. Ibid.
9. *The Times*, 20 Mar. 1876, 12a.
10. *HL Debs*, 27 Mar. 1876, col. 600.
11. Ibid., col. 602.
12. Quoted MacLeod, R.M., 'The Alkali Acts administration, 1863-84 ...', *Victorian Studies*, IX, No. 2 97 (Dec. 1965).
13. Cf. memorial to Prest of Local Govt Bd, Sep. 1874; MH 16/1.
14. Ref. 12, p. 94.
15. To John Lambert, Permanent Secy, Local Govt Bd, 1 Feb. 1875; MH 16/1.
16. Ibid.
17. To Lambert, 24 Feb. 1875; ibid.
18. 2 Mar. 1875; ibid.
19. Lambert to Prest, 14 Dec. 1875; MH 16/1.
20. Memo., 16 April 1883; MH 16/2.
21. *HL Debs*, 15 Feb. 1881, col. 871.
22. *Parl. Pp.* (*HC*) 1878 (C 2195-I) XLIV, q. 9543.
23. Ibid., q. 9440.
24. Ibid., qs 9615, 9624.
25. Ibid., q. 9615.
26. *Report*, p. 15; *Parl. Pp.* (*HC*) 1878 (C2159) XLIV.
27. Ibid., p. 16.
28. Ibid., p. 3.
29. Ref. 27.
30. Ref. 22, q. 356.
31. Ibid., q. 194; cf. qs 12180, 12245.
32. Ref. 6.
33. Ref. 22, q. 6590.
34. Ibid., q. 355.
35. Ref. 33.
36. Ibid., q. 6595.
37. Ibid., q. 7066.
38. Ibid., q. 7227.
39. Ibid., q. 81.
40. Ibid., qs 86ff., 6819.
41. Ibid., q. 6260.
42. Ibid., qs 4589, 7029.
43. Cf. Lambert, ibid., q. 13155.
44. Ibid., q. 12965.
45. Ibid., q. 13204.
46. Ibid., q. 13225.
47. Ibid., qs 12882 ff., 13200 ff.
48. Ref. 26, p. 30.
49. Ibid., pp. 30-1.
50. Ibid., pp. 32-3.

Chapter 4

1. *The Times*, 22 Nov. 1878, 7d.
2. Ibid.
3. Ibid., 6 Feb. 1879, 6e.
4. MH 16/1.
5. *The Times*, 13 Feb. 1880, 7f.
6. *Parl. Pp.* (*HC*) 1880 (74) I.
7. *The Times*, 10 May 1880, 8c.
8. *HL Debs*, 4 Feb. 1881, col. 158.
9. Ibid., 15 Feb. 1881, col. 867.
10. Ibid., cols 873-4.
11. *The Times*, 17 Feb. 1881, 9e.
12. 22 Feb. 1881; MH 16/1.
13. *HC Debs*, 2 May 1881, col. 1629 ff.
14. Ibid., 13 June 1881, col. 447.
15. 44 & 45 Vict. c. 37.
16. 12 Nov. 1880; MH 16/1.
17. 3 June 1880; ibid.
18. 11 Mar. 1881; ibid.
19. Ibid.
20. 21 Mar. 1881; ibid.
21. To Secy, Local Govt Bd, 25 Mar. 1881; ibid.
22. Minute by Prest, Local Govt Bd, 29 Mar. 1881; ibid.
23. *AR for 1879* (1880) 20.
24. Ref. 16.
25. p. 40 above.
26. Cl. 4, Alkali &c Works Regulation Bill (No 119) 18 Mar. 1881; *Parl. Pp.* (*HC*) 1881, I.
27. Estimates for civil service and revenue departments for the year ending 31 Mar. 1883; *Parl. Pp.* (*HC*) 1882, XLII.
28. Fletcher to Prest, Local Govt Bd, 20 Jan. 1880; Smith to Secy, 8 Dec. 1880; MH 16/1.
29. *AR for 1881*, Aug. 1882 (1883).
30. District report, quoted *AR for 1882* (1883).
31. Smith to Local Govt Bd, 20 Feb. 1883; MH 16/2.
32. See below p. 129.
33. 8 Dec. 1880; MH 16/1.
34. See pp. 130-4.
35. *AR for 1871* (1872) 5.
36. *AR for 1881*, Aug. 1882 (1883).
37. To Local Govt Bd, 16 Apr. 1883; MH 16/2; *AR for 1882* (1883).
38. To Local Govt Bd, 21 Jan. 1884; MH 16/2.
39. To Secy, Local Govt Bd, 8 Dec. 1880; MH 16/1.
40. To Local Govt Bd, 20 Feb. 1883, MH 16/2.

Chapter 5

1. p. 52 above.
2. The Local Government Act, 1858; 21 & 22 Vict. c. 98.
3. The Sanitary Act, 1866; 29 & 30 Vict. c. 90.

4. 38 & 39 Vict. c. 55.
5. Simon, J., *English sanitary institutions.* p. 467. Cassell & Co., London (1890).
6. Enclosure in A.C. Bruce, Asst Commr Met. Police to HO, 16 Feb. 1887; HO 45/9991/A 46381.
7. *The Times*, 5 Nov. 1880, 8e; 13 Nov. 1883, 2f.
8. Russell, F.A.R., *London fogs*. Stanford, London (1880).
9. Russell, F.A.R., *Smoke in relation to fogs in London*. National Smoke Abatement Offices, London [1889].
10. *The Times*, 13 Mar. 1882, 12b.
11. p. 7 above.
12. *The Times*, 25 Oct. 1880, 10e.
13. Ibid., 18 Nov. 1880, 6d.
14. Ibid., 10 Jan. 1881, 4f.
15. Ibid., 21 Feb. 1881, 8a.
16. Ref. 14.
17. *The Times*, 5 Nov. 1880, 8e.
18. Ibid., 1 Dec. 1881, 11f.
19. Ibid., 19 Jan. 1882, 10f; 6 Feb. 1882, 7e.
20. Smoke Abatement Committee, *Report...1882, with reports of the Exhibition at South Kensington...* Smith, Elder & Co., London (1893).
21. *The Times*, 18 July 1883, 6f.
22. Ibid.
23. *Parl. Pp.* (*HL*) 1884 (109) XI.
24. *HL Debs*, 26 May 1884, col. 1267.
25. Ibid., 31 July 1884, col. 1129 ff.
26. *Parl. Pp.* (*HL*) 1884-85 (50) VI.
27. Ibid., 1886 (27) VI.
28. Secy, NSAI, to Home Secy, 7 Feb. 1887; HO 45/9991/A 46381.
29. Minute, 24 Mar. 1887; ibid.
30. Undated, ibid.
31. To Secy, NSAI, 6 Apr. 1887; ibid.
32. *Parl. Pp.* (*HL*) 1887 (43) II.
33. *HL Debs*, 20 June 1887, col. 531 ff.
34. *Parl. Pp.* (*HL*) 1887 (174) III.
35. *HL Debs*, 1 Aug. 1887, col. 676 ff.
36. *Parl. Pp.* (*HL*) 1888 (64) VIII.
37. Ibid., 1889 (13) VI.
38. Ibid., 1889 (245) VI.
39. Ibid., 1890 (45) VIII.
40. Ibid., 1891 (22) VI.
41. Ibid., 1892 (9) VIII.
42. *HL Debs*, 2 Mar. 1891, col. 1906.
43. p. 7 above.
44. *HL Debs*, 12 Feb. 1892, col. 301 ff.
45. *HC Debs*, 6 June 1893, col. 327.
46. *The Times*, 13 Feb. 1892, 9c.
47. 'Smoke abatement exhibition', *Nature* **25** 219 (1882).
48. *The Times*, 23 May 1890, 9d.

Chapter 6

1. *The Times*, 13 May 1884, 10d.
2. MH 16/2.
3. Obituary notice, *The Times*, 22 Sep. 1920.
4. *AR for 1887* (1888) 15 ff.
5. p. 40 above.
6. To Secy, Local Govt Bd, 24 Jan. 1889; MH 16/3.
7. To the President, 2 Feb. 1889; ibid.
8. 15 Apr. 1889; ibid.
9. 11 May 1889; ibid.
10. From the Barton Regis Union Rural Sanitary Authority, 22 Nov. 1889; from Fletcher, 25 Nov. 1889; ibid.
11. 16 Dec. 1889; MH 16/3.
12. Minutes, 18 and 21 Dec. 1889; ibid.
13. 23 Jan. 1890; ibid.
14. 30 Jan. 1890; ibid.
15. 'Proposed amendment of the Alkali etc., Works Regulation Act 1881' 24 Feb. 1890; ibid.
16. 11 Mar. 1890; ibid.
17. 5 May 1890; ibid.
18. Minute of President, 13 Apr. 1891; ibid.
19. 22 June 1891; ibid.
20. *AR for 1890* (1891) 23.
21. 20 Oct. 1891; MH 16/3.
22. 27 Oct. 1891; ibid.
23. Minutes, 4 Nov. 1891; ibid.
24. 30 Dec. 1891; ibid.
25. 4 Mar. 1892; ibid.
26. Alkali, &c. Works Regulation Act, 1892, 55 & 56 Vict. c. 30.
27. Memorial to the Local Govt Bd, 14 June 1892; MH 16/3.
28. *AR* (*Scotland*) *for 1892* (1893) 115; *AR for 1893* (1894) 32.
29. p. 40 above.
30. *AR for 1884* (1885) 11 ff.
31. *AR for 1885* (1886) 8.
32. *AR for 1886* (1887) 13.
33. *AR* (*Scotland*) *for 1888* (1889) 112.
34. *Parl. Pp.* (*HC*) 1878 (C 2159-I) XLIV, q. 6595.
35. Cf. *AR for 1887* (1888).
36. *AR for 1891* (1892) 8.
37. Cf. evidence to Aberdare Commission; ref. 34, qs 6743, 6859.
38. Provisional Order Confirmation (Salt Works) Act, 1884; 47 & 48 Vict. c. 157.
39. *AR for 1885* (1886) 13.
40. p. 58 above.
41. *AR for 1886* (1887) 18 ff.
42. *The Times*, 3 Aug. 1896, 10c.
43. *AR for 1890* (1891) 26 ff.
44. *AR for 1894* (1895) 21 ff.
45. *AR for 1889* (1890).
46. *AR* (*Scotland*) *for 1889* (1890) 97; *AR for 1894* (1895).
47. *AR for 1889* (1890) 12.
48. Appeals of 27 Mar. 1892, 5 Jan. 1893 and 19 Mar. 1894; MH 16/3 and MH 16/4.

49. 10 Mar. 1893; MH 16/4.
50. To Secy, 6 Aug. 1890; from him, 19 Sept. 1890; MH 16/3.
51. 18 Aug. 1890; ibid.
52. *AR for 1885* (1886) 8.
53. Cf. the reminiscences of the Chief Inspector for Scotland, E.A. Balfour Birse, in his report for the centenary year of the first Alkali Act; *AR (Scotland) for 1963* (1964) Appx VI.
54. Cf. the tribute of T.L. Bailey, a later Chief Inspector, in his annual report for 1920; *AR for 1920* (1921).
55. He died on 15 Sept. 1920, and *The Times* carried an obituary notice of him a week later; ref. 3.

Chapter 7

1. W.S. Curphey's testimony in recording his death on 1 Feb. 1915; *AR for 1914* (1915).
2. *AR for 1899* (1900) 10.
3. From the Local Govt Bd, 19 June 1896; MH 16/4.
4. Minute to the Secy, recd 26 Nov. 1898; ibid.
5. Ibid.
6. The Bill (Alkali &c Works Regulation Bill [HL]) was given a 3rd reading in the HL on 20 May 1901, and introduced into the HC on 11 June 1901; *Parl. Pp.* (*HC*) 1901 (207) I.
7. *HC Debs*, 22 July 1901, col. 1223.
8. Ibid., col. 1229.
9. Ibid., col. 1230.
10. *Parl. Pp.* (*HC*) 1903 (325) I.
11. Ibid., 1904 (202) I.
12. Ibid., 1905 (227) I.
13. Ibid., 1906 (109) I.
14. 6 Edw.7 c.14.
15. Ibid., sect. 27.
16. MH 16/4; *AR for 1905*, 17 May 1906 (1906) and *for 1909*, 18 Mar. 1910 (1910); *AR for 1910* (1911).
17. *Parl. Pp.* (*HC*) 1878 (C 2159-I) XLIV, q. 11487.
18. p. 64 above.
19. *The Times*, 19 Nov. 1898, 11d.
20. Ibid., 18 Nov. 1898, 8d.
21. 54 & 55 Vict. c.76.
22. Ref. 19.
23. Ref. 21, sect. 100.
24. *The Times*, 24 Nov. 1898, 12b.
25. Ibid.
26. Ibid., 6 Dec. 1898, 10e.
27. Ibid., 22 Nov. 1899, 11f.
28. Ibid., 23 Apr. 1907, 2f.
29. Ibid., 20 Nov. 1905, 6e.
30. Chief Officer, Public Control Dept, LCC, 'Smoke nuisance in London' [30 Nov. 1906]; C/c Misc. Pp. 799 (Greater London Record Office).
31. Ref. 28.
32. *The Times*, 6 Mar. 1899, 10f.

33. Ibid., 18 Dec. 1905, 9c.
34. Ibid., 27 Dec. 1910, 11a.
35. LCC to Home Office, 19 Dec. 1906, Home Office to LCC, 15 Jan. 1907; HO 45/9991/A 46381.
36. *The Times*, 4 Nov. 1907, 12f.
37. Ref. 30.
38. Cf. p. 83 above.
39. LCC, C/c Misc. Pp. 799 (GLRO).
40. LCC Parliamentary Committee, Memorandum for the Public Control Committee, 17 Feb. 1910; ibid.
41. LCC Parliamentary Committee, Report on the LCC (General Powers) Bill, 1910, 7 July 1910; ibid.
42. LCC Public Control Committee, Minutes, 18 Feb. 1910; ibid.
43. Cf. reported proceedings, *The Times*, 13-16 Dec. 1905: 10e, 15c, 7c, 7b.
44. Ibid., 14 Dec. 1905, 15c.
45. Ibid., evidence of Prof. Smithells to Newton Committee, 16 Mar. 1920, HLG 55/42; report of Newton Committee, para.6, Ministry of Health, Committee on smoke and noxious vapours abatement, *Final Report*. HMSO, London (1921).
46. *Times Engineering Supplement*, 22 Sep. 1909, 17e.
47. *The Times*, 30 June 1910, 8c.
48. *Newton Report* (ref. 45) para. 14.
49. *Times Engineering Supplement*, 27 Mar. 1912, 26b.
50. *The Times*, 5 Jan. 1912, 5e; *Times Engineering Supplement*, 10 Jan. 1912, 20b.
51. *HC Debs*, 30 Apr. 1913, col. 1214; *Parl. Pp.* (*HC*) 1913 (136) V.
52. *HL Debs*, 17 Mar. 1914, col. 502; *Parl. Pp.* (*HL*) 1914 (30) V.
53. *Times Engineering Supplement*, 7 Jan. 1914, 3b.
54. *HL Debs*, 24 Mar. 1914, cols 665, 666.
55. Ibid., col. 671.
56. *AR for 1912* (1913).
57. *AR for 1911* (1912).
58. *Ars for 1915-18* (1916-19).
59. *AR for 1918* (1919).

Chapter 8

1. *The Times*, 4 Mar. 1919, 6c.
2. *HL Debs*, 24 Mar. 1914, cols. 665-6.
3. Ministry of Health, Committee on smoke and noxious vapours abatement, *Final report.* HMSO, London (1921).
4. Ibid., paras. 106-10.
5. Ibid., paras. 87, 80-5.
6. Ibid., para. 93.
7. Ibid., paras. 96-7.
8. *Parl. Pp.* (*HC*) 1920 (Cmd 755) XXV. Reprinted in ref. 3.
9. Ref. 3, Appx A, 33.
10. 9 Jan. 1922; HLG 55/17.
11. *The Times*, 21 Mar. 1922, 9a.
12. *HL Debs*, 10 May 1922, col. 374.
13. *Parl. Pp.* (*HL*) 1922 (168) IV.

14. *The Times*, 26 Mar. 1923, 7f.
15. *Parl. Pp.* (*HL*) 1926 (51) IV.
16. Alkali &c Works Order, 1928.
17. *HL Debs*, 23 Mar. 1926, cols. 719, 722.
18. *Smoke Abatement Bill, Standing Cttee B, rept and procs*; 1926 (147) VII 4,5.
19. *HC Debs*, 22 June 1926, cols. 288-9.
20. Ibid., 6 Dec. 1926, col. 1837.
21. Ibid., col. 1838.
22. Ministry of Health memorandum, enclosed in minute to Secy of 15 Jan. 1932; HLG 55/8.
23. Ministry of Health review of work on smoke abatement enclosed in minute of 7 Nov. 1932; ibid.
24. Ministry of Fuel and Power, Fuel and Power Advisory Council, *Domestic fuel policy*, Cmd 6762. HMSO, London (1946).
25. Ibid., appx.II, 45.
26. *Smokeless Air*, No.66 29 (1948).
27. Ref. 10.
28. Ibid.
29. Private information; letter of 4 June 1978.
30. *AR for 1959* (1960).
31. Report on 'Atmospheric pollution', 9 Aug. 1932; ref. 23, appx D.
32. p. 98 above.
33. W.A. Ross to I.G. Gibbon, 7 Nov. 1932; HLG 55/8.
34. pp. 1 and 55 above.
35. *AR for 1936* (1937).
36. Ibid., cf. p. 8 above.
37. *AR for 1935* (1936) 3.
38. *AR for 1948* (1949).
39. He wrote in his annual report for 1946: 'The Minister is usually willing that the Alkali Inspectors should investigate any case of genuine complaint but it should be understood that the Alkali Inspectors' authority is confined to works registered under the Alkali Act and that they cannot impose conditions on works which are not so registrable.' *AR for 1946*, 18 Feb. 1947 (1948).
40. *Chemistry & Industry*, Apr. 1949.
41. *AR for 1949* (1950); Alkali &c Works Order, 1950 (Statutory Instruments 1950, No 364).
42. *AR for 1920* (1921).
43. *AR for 1937* (1938).
44. *AR for 1946* (1948).
45. *AR for 1938* (1939).
46. *AR for 1939-1945* (1946).
47. *AR for 1948* (1949).
48. *AR for 1951* (1952).
49. Ref. 40.
50. Ibid., Feb. 1947.

Chapter 9

1. NSAS, *Annual report, 1953*, quoted Sanderson, J.B., 'The National Smoke Abatement Society and the Clean Air Act (1956)', *Political Studies*, IX 236-53 (1961).

2. *The Times*, 8 Dec. 1952.
3. Ministry of Health, *Mortality and morbidity during the London fog of December 1952*. HMSO, London (1954).
4. *The Times*, 9 Dec. 1952.
5. *HC Debs*, 18 Dec. 1952.
6. *The Times*, 31 Jan. 1953.
7. p. 55 above.
8. *HC Debs*, 27 Jan. 1953, col. 829.
9. Ibid., col. 831.
10. p. 99 above.
11. *HC Debs*, 13 Mar. 1953 col. *155*.
12. *Evening Standard*, 24 Jan. 1953, quoted *HC Debs*, 8 May 1953, col.845.
13. *HC Debs*, 31 July 1953, col. 201.
14. Private information.
15. Committee on air pollution, *Interim report*, Cmd 9011. HMSO, London (1953); *Report*, Cmd 9322. HMSO, London (1954).
16. Ibid., para. 6.
17. Ibid., paras. 99-100.
18. Ibid., para. 118.
19. Ibid., para. 76.
20. *HC Debs*, 25 Jan. 1955, col. 87.
21. Ibid., 10 Mar. 1954, col. 2306 ff.
22. Quoted, Foulkes, K., *The background and terms of the Clean Air Act 1956*. Unpubd thesis, University of Sussex (1971).
23. *HC Debs*, 4 Feb. 1955, col. 1466.
24. Ibid., col. 1479.
25. Ibid., col. 1488 ff.
26. Ibid., 3 Nov. 1955, col. 1247 ff.
27. Ref. 15, para. 45.
28. *HC Debs*, 3 Nov. 1955, col. 1328 ff.
29. *HL Debs*, 30 May 1956, col. 644.
30. Clean Air Act, 1956, 4 & 5 Eliz.2 c.52, sect. 34.
31. Ref. 29, col. 594.
32. For a detailed treatment of this topic see Foster, L.T., *The response of local government to the U.K. Clean Air Act, 1956: a case study in the adoption of permissive environment legislation.* Unpubd thesis, University of Toronto (1975).
33. Bill to make further provision for abating the pollution of the air. Bill No 27.
34. *HC Debs*, 2 Feb. 1968, cols. 1801-8.
35. Elizabeth II, 1968, c.62.
36. Royal Commission on Environmental Pollution, *First report*, Cmnd 4585. HMSO, London (1971).

Chapter 10

1. *AR for 1960* (1961) 7.
2. *Alkali &c Works Orders, 1966 and 1971;* Statutory Instruments 1971, No 960.
3. *AR for 1958* (1959).
4. Ibid.
5. *AR for 1959* (1960).
6. *AR* (*Scotland*) *for 1958* (1959).

7. *HC Debs*, 13 May 1958, col. 359.
8. Ibid., 29 Jan. 1962, col. 837.
9. *AR for 1958* (1959).
10. *AR for 1959* (1960) 41.
11. *AR for 1961* (1962) 29.
12. *AR for 1962* (1963) 32.
13. *AR for 1963* (1964) 2.
14. Bugler, Jeremy, *Polluting Britain: a report*. Penguin, London (1972); Clean Air Council, 44th meeting; CACP 353/2 Pt 1 (Department of Environment records).
15. *HC Debs*, 28 Feb. 1973, col. 1502; *Parl. Pp.* (*HC*) 1972-73 (81) I.
16. *HC Debs*, 11 May 1973, col. 942 ff.
17. *AR for 1968* (1969) 7.
18. Committee on Safety and Health at Work, *Report*, Cmnd 5034. HMSO, London (1972).
19. Ibid., para. 139.
20. Ibid., para. 193.
21. Royal Commission on Environmental Pollution, *Fifth report; air pollution control: an integrated approach*, Cmnd 6371. HMSO, London (1976) paras. 257, 260.
22. Frankel, Maurice, *The Alkali Inspectorate: the control of industrial air pollution.* (Social Audit special report) Social Audit Ltd, London (1974).
23. pp. 142, 145.
24. Ref. 21.
25. Ibid., para. 16.
26. Ibid., para. 271.
27. *Parl. Pp.* (*HC*) 1878 (C 2159-I) XLIV, q. 8610.
28. Ibid., q. 8637.
29. Ibid., q. 10063.
30. Ibid., q. 12138.
31. Alkali &c Works Regulation Act, 1881, sect. 10.
32. Local Government Provisional Order (Salt Works and Cement) Bill, No 216, introduced in the HC on 19 May 1884; cf. *AR for 1884* (1885) 13.
33. Alkali &c Works Order, 1935.
34. *AR for 1960* (1961).

Chapter 11

1. Frankel, Maurice, see ref. 22, Chapter 10.
2. *AR for 1965* (1966).
3. *AR for 1971* (1972) 12.
4. *AR for 1970* (1971) 5-6.
5. Royal Commission on Environmental Pollution, *Second report: three issues in industrial pollution*, Cmnd 4894. HMSO, London (1972).
6. Department of the Environment, *Information about industrial emissions to the atmosphere*. Report by a working party of the Clean Air Council. HMSO, London (1873).
7. Royal Commission on Electric Power Planning (Ontario), I. *Concepts, conclusions, and recommendations.* Toronto (1980).
8. Tunnicliffe, M.F., The interpretation of ''best practicable means''. *The Chemical Engineer* **271** 121-4 (1973).

9. Health and Safety Executive, *Industrial air pollution, 1975*. pp. 54-60. HMSO, London (1977).
10. Auliciems, A. and Burton, I., Trends in smoke concentration before and after the Clean Air Act of 1956. *Atmospheric Environment* **7** 1063-70 (1973).
11. Department of Trade and Industry, Warren Spring Laboratory, *National survey of air pollution, 1961-71*. HMSO, London (1972).
12. Department of the Environment, Working party on grit and dust emissions, *Report*. HMSO, London (1974).
13. Department of the Environment, Working party on the suppression of odours from offensive and selected other trades, *Reports, part i and part ii*. HMSO, London (1974 and 1975).
14. Department of the Environment, *Pollution papers* **2** (1974), **9** (1976), **12** (1977), **14** (1978), **15** (1979). HMSO London.
15. Ibid., *Pollution paper 4: controlling pollution*. HMSO, London (1975).
16. Council on Environmental Quality, *Environmental quality: 10th annual report*. pp. 666-7. Washington, DC (1979).
17. Ibid. *Environmental quality: 9th annual report*. p. 421. Washington, DC (1978).
18. Programmes Analysis Unit, *An economic and technical appraisal of air pollution in the United Kingdom*. HMSO, London (1972).
19. Royal College of Physicians, *Air pollution and health*. Pitman, London (1970).
20. Douglas, J.W.E. and Waller, R.E., Air pollution and respiratory infection in children. *British Journal of Preventive and Social Medicine*, **20** 1-8 (1966).
21. Examples of recent books on this subject are: Kates, R.W., *Risk assessment of environmental hazard*. Wiley, Chichester, (1978); Lowrence, W.W., *Of acceptable risk*. Kaufman, Los Altos, Cal. (1976); Goodman, G.T. and Rowe, W.D. (Eds), *Energy risk management*. Academic Press, London (1979); Schwing, R.C. and Albers, W.A. (Eds), *Society risk assessment*. Plenum Press, New York (1980).
22. Ashby, E., *Reconciling man with the environment*. Stanford University Press (1978).
23. Lundholm, B., Interactions between oceans and terrestrial systems. In *Global effects of environmental pollution* (ed. S.F. Singer), pp. 195-201. Reidel, Dordrecht, Holland (1970).
24. Royal Ministry for Foreign Affairs and Royal Ministry of Agriculture, Sweden *Air pollution across national boundaries*. Stockholm (1971).
25. OECD Environmental Directorate. *The OECD programme on long range transport of air pollutants: summary report*. OECD, Paris (1977).
26. Scriven, R.A. and Howells, G., Stack emissions and the environment, *CEBG Research*, No. 5, 28-40 (1977).
27. *Treaty establishing the European Economic Community, Rome, 25 March, 1957*, Cmnd. 4864. HMSO, London (1972).
28. Declaration of the Council of the European Communities...on the programme of action of the European Communities on the environment. *Official Journal*, 16 (20 Dec. 1973) C-112.
29. Proposal for a Council Directive on the use of fuel oils with the aim of decreasing sulphurous emissions. *Official Journal*, 19 (8 Mar. 1976) C-54.

INDEX

Abbreviations

CSAS Coal Smoke Abatement Society
NSAI National Smoke Abatement Institution
NSAS National Smoke Abatement Society
NSCA National Society for Clean Air